21st Century Electricity

By

D J Haskell

Published by New Generation Publishing in 2015

Copyright © D J Haskell 2015

First Edition

The author asserts the moral right under the Copyright, Designs and Patents Act 1988 to be identified as the author of this work.

All Rights reserved. No part of this publication may be reproduced, stored in a retrieval system or transmitted, in any form or by any means without the prior consent of the author, nor be otherwise circulated in any form of binding or cover other than that which it is published and without a similar condition being imposed on the subsequent purchaser. All opinions expressed in this book, unless otherwise stated, are purely those of the author.

Liability Notice

While every effort has been taken with the researching, writing and production of this book to ensure the accuracy of the information contained herein, the Publishers and Author do not in any way accept responsibility or liability for any such erroneous statements or claims herein and/or any loss or damage relating thereof as a direct or indirect consequence of the information contained within this book.

www.newgeneration-publishing.com

The Metric and Imperial System of Measurement.

Although the Metric system is used throughout the book there are a number of places where the Imperial System of measurement is also shown in brackets. The author recognises there are many people who, like himself, have 'grown up' with both systems. The author whilst having an appreciation of metric measurements still tends to think and to be more at ease with the scales of distance and size in Imperial Units. As an example the author has a better grasp for the meaning of five hundred miles as against eight hundred kilometres, or twenty-two pounds rather than ten kilograms.

Acknowledgements

Dr John Etherington PhD DIC BSc ARCS for his generous patience in answering my many queries relating to large scale wind generation in the UK. I would also highly recommend his excellent book titled, The Wind Farm Scam, ISBN: 978 1905299 836.

Angela Kelly of Country Guardian, (www.countryguardian.net), a true guardian of the countryside.

Mark Duchamp, President, Save the Eagles International, whose tireless efforts know no bounds. (www.savetheeaglesinternational.org).

Lemnis Lighting, The Netherlands, for the kind provision of their Pharox lamp for evaluation. (www.lemnislighting.com).

Solar iBoost, Marlec Engineering Co Ltd, Rutland House, Trevithick Road, Corby, Northants, NN17 5XY. (www.marlec.co.uk).

Flow Energy, Felaw Maltings, 48 Felaw Street, Ipswich, IP2 8PN. (www.flowenergy.uk.com).

Christopher Booker, The Sunday Telegraph, 23 March 2013.

Geological Society of London. (www.geolsoc.org.uk/shalegas).

Tidal Electric, 50 Albermarle Street, London, W1S 4BD. (www.tidalelectric.com).

Blue Energy Canada Inc. Canada. (www.bluenergy.com).

Wavegen (A Voith and Siemens Company), Inverness, Scotland.

Bat Conservation Trust (BCT). (www.bats.org.uk/pages/wind_turbines.html).

Astronomy Now magazine.

British Astronomical Association's Campaign for Dark Skies.

Marine Current Turbines, The Court, The Green, Stoke Gifford, Bristol, BS34 8PD. (www.marineturbines.com).

Daily Mail, Northcliffe House, 2 Derry Street, London, W8 5TT, (www.dailymail.co.uk).

Ceres Power Limited, Unit 18, Denvale Trade Park, Haslett Avenue East, Crawley, RH10 1SS.

Sustainable Development Commission (SDC), Ground Floor, Ergon House, Horseferry Road, London, SW1P 2AL.

Windjenpower Ltd, Colwyn Bay, Conwy, LL29 8TH.

World Energy Council, 5th Floor, Regency House, 1-4 Warwick Street, London, W1B 5LT.

National Grid, National Grid House, Warwick Technology Park, Warwick, CV34 6DA.

The Department for Business, Enterprise and Regulatory Reform (BERR) formerly DTI.

British Nuclear Fuels (BNFL).

National lightning Safety Institute, 891 N. Hoover Avenue, Louisville CO 80027, USA.

European Wind Energy Association (EWEA).

British Wind Energy Association (BWEA).

E.ON, German Power Company (formerly Powergen).

Ofgem, 9 Millbank, London, SW1P 3GE.

National Statistics for Council tax dwellings Wales. (www.wales.gov.uk/statistics).

Unstoppable Global Warming by S. Fred Singer and Dennis T. Avery.

The Chilling Stars by Henril Svensmark and Nigel Calder.

Love Food Hate Waste is brought to you by WRAP (www.wrap.org.uk).

Environmental Protection UK (www.environmental-protection.org.uk).

Energy Saving Trust (www.energysavingtrust.org.uk).

Green Frog power Ltd, Birmingham (wwww.greenfrogpower.co.uk).

YorPower, Yorkshire (www.yorpower.com).

PowerSines Ltd (www.powersines.com).

Electrical Safety First (www.electricalsafetyfirst.org.uk).

Smith's Environmental Products Ltd, Blackall Industrial Estate, South Woodham Ferrers, Chelmsford, Essex. (www.smiths-env.com).

Worcester Bosch Group, Cotswold Way, Warndon, Worcester, WR4 9SW. (www.worcester-bosch.co.uk).

The numerous journals, magazines, books and electronic media I have read that have contributed to my knowledge during research into this subject.

Last, but not least, my wife Patricia for her patience, understanding and encouragement whilst I was researching and writing this book.

Foreword

By David Bellamy, Bedburn, March 2014.

Everything the Wind Energy Associations never wanted to see in print!

The scientific facts and truths about the inadequacies of wind power and the horrendous cost both to the consumer and to our lives and countryside.

We do need sources of alternative energy but wind does not even deserve that name because as this timely exposé shows it is nothing but a very costly add-on extra that can do little or nothing to slow climate change. Neither can it give us security of electricity supply without continuous back up from fossil fuel fired power stations. Read all about it and then join the thousands of people who are campaigning across the British Isles against these wind monsters.

Preface

If industrial, commercial and domestic electrical energy should enigmatically disappear overnight it would prove to be cataclysmic for our civilised society - make no mistake, modern civilisation would totally collapse and if people did survive the breakdown of society then they would have to eke out an existence in dark, damp and cold buildings – apart from lighting, central heating needs electricity for system control and pump operation. Millions of people in a collapsed civilised world would eventually starve due to the failure of various vital systems – supermarkets would soon be depleted of all food, drink and other goods – transportation would grind to a halt from eventual lack of fuel – hospitals would stop fully functioning - people would suffer from illness, epidemics leading to pandemics - eventually perishing in large numbers.

It can be argued that only primitive tribes not dependent on such electrical energy would survive, living as they do now following a primitive means of existence - if this nightmare scenario should actually materialise then all may not be totally lost as given some luck and the necessary time, Homo sapien might eventually again improve his lot and eventually start again to mirror ancient civilisations such as the Chinese, Sumerian, Egyptian, Greek or Roman - thus starting the whole ball rolling again…

It is all too easy to overlook how dependent we have become on electricity for our very existence ranging from mobile telephones, personal computers, transport, terrestrial and satellite communication, radio and television, the supply and running of super markets and vehicle filling stations, the functioning of oil refineries, hospitals, factories, offices, world banking to the military – think of all the things that are dependent on electricity and then imagine the civilised world suddenly without access to this now vital energy – a frightening thought!

A loss of electricity would mean that high technological efforts such as the Mars Atmosphere and Volatile EvolutioN (MAVEN) orbital mission will be lost and possibly end its days in a Martian grave. But then what of the International Space Station (ISS) and the astronauts on-board - how would they communicate with Earth and indeed return to the surface of our planet – even if they did land safely then what kind of world will they be returning to?

The above scenario may seem somewhat far-fetched - but how fanciful is it - when you consider the inept situation relating to UK energy policy brought about by ill-informed politicians - how long it will be before the lights begin to go out – such follies as wind farms and solar parks will definitely not save us.

It is this incompetence in conjunction with other realisations, such as Solar Corona Mass Ejections (CMEs), the avarice, ignorance and the myopic thinking of our ruling class that motivated the writing of this book - to enlighten the general reader to the current state of power generation in the UK – the lack of any coherent plans for the future, and the ill-conceived and extremely costly concept and on-going provision of large scale wind electrical generation in the United Kingdom.

During my early research it soon became apparent I was being somewhat naïve and limited in my approach and proposed scope of the subject. The initial plan was to concentrate purely on the history and distribution of wind generated electricity, although I rapidly became aware of the delusions, misunderstanding and indeed ignorance a lot of folk have regarding electrical energy, its generation, distribution and actual consumption, and unfortunately how this ignorance is being taken advantage of by unscrupulous and immoral people.

Not only is there widespread misunderstanding to the workings of the power industry, but also that of climate change, to the extent of a great falsification such that if the UK pursues its present policy with wind generation, it will have some meaningful impact on global emissions - this is the great confidence trick of the 21st Century – realising, of course, that without carbon dioxide there would be no life on this planet as we know it.

But even if additional carbon dioxide were a threat (which it isn't) then it should be very clear that if the whole of the UK should suddenly disappear under the sea then 98 per cent of emissions would still have to be addressed, as the UK contribution is just under 2 per cent of global emissions.

It is sobering to realise there are about 1,200 coal-fuelled power stations currently (2014) in the pipeline in developing countries, so it is difficult to see how even the EU can have any significant impact on the global figure when it accounts for just about 12 per cent of total emissions. Such is the madness of it all.

Terms such as wind turbine and wind farm as used by the wind industry, the media, politicians and by folk in general are both erroneous and misleading – the term wind turbine is a clever marketing ploy that conjures up the image of a much more sophisticated and powerful machine than it actually is – both terms wind turbine and wind farm are correctly addressed later in the book in the chapter headed, 'Wind Technology in the UK'.

Therefore it was deemed necessary to widen the subject matter and explain in straightforward terms an understanding of electricity, how power is distributed, and what is meant by:

- ❖ Weather.

- ❖ Climate.

- ❖ The Greenhouse Effect.

- ❖ Anthropogenic Global Warming (AGW).

Also,

- ❖ How the Grid would cope with a giant 'solar storm' similar to the Carrington Storm during 1859.

- ❖ The limitations of wind generated electricity.

- ❖ Alternatives for power generation.

- ❖ How to cut household energy consumption – this insight alone will return handsomely the cost of this book - something every householder will want to achieve in this age of ever increasing energy prices.

It is truly amazing how many people consider a large collection of wind generators (wind turbines), commonly referred to as a wind farm to be equivalent to that of a conventional power station in the ability to provide the same magnitude and reliability of electrical power. This, of course, is absolute nonsense to anyone who understands the technology and engineering – although the limitations of wind generated electricity should not come as any surprise when considering the unpredictability and unreliability of the wind – conventional power stations provide electricity at the touch of a switch whereby wind generated electricity is totally reliant on the right kind of wind – little or no wind, or indeed too high a wind means no electricity.

Surprisingly, as mentioned above, even in the 21st Century, a large proportion of the general public have little or no understanding of power generation, or indeed what exactly electricity is. We are all guilty of ignorance to a certain degree and also of needless waste, and who will forgive our selfish despoliation of the planet - surely future generations will condemn us for our insatiable greed, this rapacity compounding our myopic and cavalier approach to many things.

It is the author's desire to 'open doors' and therefore enlighten - offering a much broader spectrum than originally intended. Huge national savings can be made in the intelligent production and usage of electricity - but only if government, business and industry has the vision and the will to execute such measures.

With regard to our decision makers my level of confidence is not high for their ability to grasp the nettle and make the right decisions - especially when considering, as an example, our Gas Storage Capacity, for at the time of writing (2014) the UK had only enough storage for 20 days usage compared to 70 days Italy, 92 days Germany, 103 days France and six months for the U.S. – remember a significant number of UK power stations run on gas.

This book is intended for general consumption and as such technical jargon, formulae and mathematics have been kept to a realistic minimum, the scope of which should be well within the grasp of the average reader. I can assure you that if you read this book with an honest and unbiased mind, you will surely and quickly appreciate the nakedness of the charlatans who would pursue and promote such an appalling industry as large scale wind generation in the United Kingdom – it is nothing less than a massive scam such that the Emperor has indeed a fine set of clothes.

Finally a quote (with acknowledgement) from the Sunday Telegraph: *As the snow of the coldest March since 1963 continues to fall, we learn that we have barely 48 hours' worth of stored gas left to keep us warm, and that the head of our second-largest electricity company, SSE, has warned that our generating capacity has fallen so low that we can expect power cuts to begin at any time. It seems the perfect storm is upon us. The grotesque mishandling of Britain's energy policy by the politicians of all 2 parties, as they chase their childish chimeras of CO_2-induced global* *warming and windmills, has been arguably the greatest act of political irresponsibility in our history. --Christopher Booker, The Sunday Telegraph, 23 March 2013.*

CONTENTS

Introduction 1

CHAPTER ONE A Brief History of Electricity 11

CHAPTER TWO Wind, Weather & Climate Change 30

CHAPTER THREE Wind Technology in the UK 66

CHAPTER FOUR Ocean Tides 119

CHAPTER FIVE Energy Alternatives 136

CHAPTER SIX Future Power Generation and Consumption 174

CHAPTER SEVEN How to Reduce your Domestic Energy Bill 205

Appendix 1 Wind 223

Appendix 2 Wind Generation 227

Appendix 3 Domestic Energy Savings 229

Appendix 4 Energy Statistics 246

Glossary of Terms 250

OTHER BOOKS BY THE AUTHOR 257

Introduction

Primarily the intention of this book was to highlight the nonsense of large scale wind generation in the UK, although early research showed that to put wind technology fully into perspective required a number of other related topics being addressed, and thus this book covers not only the madness of large scale wind generation, but a brief history of electricity, weather and climate change, alternative means of generation, ocean tides and a look to the future – and indeed, how to realistically reduce the domestic energy bill – an insight that will handsomely repay the cost of this book many times over.

It is not just generation alone that should concern us all, but that of energy conservation as well.

Thus a whole chapter is devoted to reducing domestic energy consumption - in relation to electrical energy if every household reduced their demand, then the need for more power stations is challenged thus possibly achieving a reduction in both size and number of power stations.

Indeed the potential savings achieved in the home and office, specifically with regard to the emergence of energy saving lighting should not be underestimated. Our need for illumination consumes something like 20-25 per cent of all electricity, and the new LED energy saving bulb such as the Pharox uses 90 per cent less power than its equivalent incandescent bulb. It reaches full illumination as soon as it is switched on - unlike the unpopular Compact Fluorescent Lamps (CFLs) - additionally there is no flicker from the Pharox bulb, which many people are sensitive to, and complain of. The Pharox is discussed in Chapter Six of this book.

As we enter the 21st Century, the need for a potentially fatal 230 voltage alternating current (AC) supply at consumer's premises is challenged. Co-generation is also discussed as well as questioning the continuing need for a synchronised, central power alternating current generating system - options are considered for the appraisal of a clean, reliable supply of electrical energy. Surely it makes more sense now to evaluate and assess the whole system from the customer end, rather than that of central generation - we should be looking at better management of all the Earth's resources.

We have come full circle in our use of electrical power, such that in the early days customers were persuaded to use more and more electricity – even night time usage was encouraged by offering cheaper electricity for night storage heaters - people were induced to fill their homes with more and more electrical goods. Nowadays we are guided and advised to reduce our electricity consumption, with the electricity companies telling us to be Green - to be more energy efficient by closing curtains at dusk to prevent heat being lost through windows, turning off lights when leaving a room for a long period, to turn off the standby mode of items such as the television, to use lower settings on central heating boilers, to wait until there is a full load for the dishwasher or washing machine, the list goes on and on to 'Save the Planet' and satisfy the 'Green Agenda'.

Although how this actually squares with the electricity companies whose job it is to sell units of electrical power to the customer and make a profit I am not at all sure - what will shareholders think if consumption and profits hit a low level – surely the fundamental business of the power companies is to sell electrical power to the customer and at a profit – the more the better. If not, then surely with everything being equal, the only way to maintain profits is to increase the cost per unit to the customer, or perhaps, to cut staff and resources to a bare minimum – this last option would not be a very wise choice for the well-being of the industry as it will diminish security of supply with the potential for brownouts and possible blackouts.

So why not a return to a much lower voltage level and localised Direct Current (DC) supply system? This is a reasonable question especially when it is recognised that all electronic items have relatively small power requirements, and actually operate on DC at a very much lower level than the potentially dangerous mains 230 volts AC we are all familiar with. If a localised low level DC system did come about, then it could be argued that there would be a considerable saving in manufacturing costs for these consumer items, as the provision of transformers and rectifying components would then not be required in such items as home computers, televisions, radio, sound systems and DVD players - a saving in material and costs to both manufacturer and customer.

Hopefully, having digested the contents of this book the reader will have a better understanding of the electrical producing and supply industry and as such will be able to make much more informed and intelligent decisions when contemplating electrical energy sources and usage. Is it beyond the wit of British innovation and engineering to re-assess high power demanding items, with the intention of developing and producing lower energy saving electrical goods such as kettles, electrical fires and indeed ovens, all working at a much lower voltage level than 230 volts. History would suggest it is well within the scope of Human expertise and endeavour – Homo sapien has landed on the Moon – truly all that is needed is the vision and the will.

Regarding renewable energy in the form of wind farms then what if the whole or a major part of our country were to be dependent on wind generated electricity, both onshore and offshore - power generation under these circumstances would be at the mercy of the wind. There would certainly be no security of supply as it is extremely difficult to predict accurately where and when the wind will blow across the UK - to help illustrate this point, during winter 2010, official figures revealed that on 30th December, an exceptionally still day, the United Kingdom's 3,000 operational wind generators produced only 0.04 per cent of the country's power - these figures alone illustrate the total ineffectiveness of large scale wind generated electrical energy.

But it is just not a lack of, or too little wind that is a problem as during September 2011 millions of pounds were paid to wind farm owners as a result of the wind blowing too strongly in Scotland – this illustrates the madness of it all. If the country were dependent on wind farms the lights will go out when the wind is not right - this is simply because wind generators have a limited window of opportunity and only become effective in generating meaningful and constant electrical energy at specific wind speeds. It is worth reiterating that when the wind is blowing vigorously, wind generators have to be shut down to avoid structural damage - this is something that must be considered for wind generators mounted on exposed hill tops, along coast lines and especially those situated out at sea (offshore) also bearing in mind that salt air and sea water are not good companions for electrical equipment. The efforts to obtain a meaningful and constant supply of electricity from wind machinery, such as to satisfy the requirement of the United Kingdom, will prove to be one of the greatest scams of the 21st Century, reminding me of childhood tales from the pen of such people as Hans Christian Anderson and his fairy tale of the 'Emperor's New Clothes'.

The use of the term Wind Turbine is, in itself, misleading for when people first harnessed the wind to drive a mill they called the whole device a wind mill; when the wind was used to pump water the device was called a wind pump. Therefore it follows, if only by convention, that using the wind to drive an electrical generator, then the whole device should simply be called a wind generator. Now it would be churlish not to acknowledge the clever and effective marketing of the wind industry - the word turbine implies a much more sophisticated and effective machine than it is – an air driven propeller in open flow is not a turbine nor is an alternating current generator as it is simply a generator; the combination of both propeller and alternating current generator is still not a turbine. Charles Parsons, the English inventor of the Parson's Steam Turbine, must truly be turning in his grave (pun intended).

A large wind generator simply consists of a three bladed, large propeller connected to, and propelling a relatively slow rotating shaft, which is connected to a gear box. The output of this gear box has a faster rotating shaft connected to an electrical generator - all of these components are housed in a Nacelle at the top of a tall tower - each single component is obviously not a turbine, nor is the combination of all of these components.

Any engineer worth his salt will explain that a turbine is a fairly complex piece of machinery consisting of numerous blades along a shaft fitted 'within a casing' i.e. an engine in which steam, water or gas is made to spin a rotating shaft by pushing on angled blades, like a fan. Turbines being among the most powerful machines with steam turbines driving generators in power stations and ship's propellers, with water turbines spinning the generators in hydroelectric power stations, gas turbines powering jet aircraft et cetera. I wonder how the wind industry would describe a turboprop aircraft whereby a turbine engine drives an aircraft propeller using a reduction gear.

To be sure, the most accurate description for the ineffective machine the wind industry employs is plainly a Wind Driven Generator (WDG) or Wind Generator (WG) for short - in the sense of a wind mill or wind pump - it is as simple as that - there is no ambiguity or any pretension to other than what it is.

Therefore it should be easy to see why the wind industry has 'cheekily' employed the word turbine, for it conveys their limited wind machinery a level of sophistication and effectiveness that is not deserved – although it is very effective marketing, to the extent that even anti-wind campaigners unwisely contribute to this marketing ploy by using the word turbine - and that is why, it can be argued, a lot of folk mistakenly think a wind farm (look at all those large wind turbines) to be equivalent to a conventional power station regarding magnitude and security of supply. This truly reminds me of the Latin, *non semper ea sunt quae videntur* - things are not always what they appear to be.

Wind farm is another misleading term and is discussed, along with wind generators, in Chapter Three of this book. Remember, there is no substitute for good marketing and that is possibly why a country such as the UK with its significant annual rainfall consumes large quantities of bottled water - it is commonly said that a good salesman could sell snow to an Eskimo, or sand to an Arab. Wind farms are not the answer to carbon emissions and will certainly not have any influence on the climate, to think or argue otherwise simply demonstrates appalling ignorance in these matters, and there are those silly enough to think the wind generator will solve our future electrical energy needs.

This book will adequately show the utter nonsense of the various nefarious claims relating to wind farms - whilst it can be argued that their wind driven operation does not directly contribute to atmospheric pollution, the construction of, conveyance to site, on-site preparation, access roads, and connection to the electricity network, does indeed use energy and contribute to pollution. It should be clearly understood that the construction and erection of wind generators for a wind farm require thousands of tons of aggregate and concrete for their necessary foundations, not to mention the material for the approach roads required for the on-going repair and maintenance of these monstrous machines - materials such as cement consume large quantities of carbon-based fuels, such as coal or oil; these fuels emitting large quantities of greenhouse gases when burnt in cement kilns during the manufacturing stage. Additionally the aggregate required in concrete making also demands carbon-based fuels for the machinery used in extraction/quarrying process.

Next, one has to consider the necessary transportation of the material and all what that entails; all these procedures contribute to considerable amounts of atmospheric pollutants such as carbon monoxide (CO) and other obnoxious emissions from vehicular exhausts. Then there is the erection of the necessary wires/cables, poles and/or pylons for connecting to the local distribution or grid transmission network – so is it surprising that wind farms are not as totally environmentally friendly as their misguided supporters would have us believe.

Nevertheless it is acknowledged that relatively small wind generators (which are defined as mini and micro turbines) can meet a requirement as it is recognised there are places around the globe and indeed in the UK where the employment of this type of generator has a justifiable case. Generally these are at places such as cattle farms in the Australian outback, places which make any connection to a Grid system practically impossible due to distance and economics; the same criteria applying to cattle ranches in the USA and other similar places.

Although petrol or diesel driven generators and solar power are available choices in themselves, who would deny these isolated places the option of intermittent wind generated electricity since they certainly do not affect anybody else! Obviously, when there is no wind or indeed too strong a wind, no electrical power is generated. So it makes sense to have a backup source of electrical energy in the form of a petrol or diesel driven generator.

Parts of the UK that could argue a case (taking into account wild life and the beauty of the landscape) for a small wind generator would be the remote islands off the coast of Scotland, where connection to the local distribution system or Grid may not be viable due to geographical placement and economics. Backup again would be required for guaranteed power, and even if the wind generator provided direct current to charge a bank of batteries, it would still be sensible to have a petrol or diesel driven generator standing by for prolonged periods of very strong, weak or no winds as a battery has a finite charge. Obviously, very small wind generators have a good use when employed on yachts and other water craft for charging batteries, not forgetting caravans and other similar small electrical energy requirements.

Having put the case for small and micro wind generators (in the right location) it should be recognised that the sighting of small wind generators on a roof of a house is a complete nonsense and a total waste of time and money, especially in urban areas and the chapter, 'Wind Technology in the UK' clarifies this in more depth.

It naturally follows that wind generators by their very nature need to be carefully planned and located in areas where there is a constant, non-fluctuating level of wind; that is, neither too weak or too strong. No one in their right mind would consider constructing a wind farm in the Doldrums for the obvious reason (apart from other factors) that there is very little or no wind there.

The reader should fully appreciate the small and limited output of an average wind farm is just not in the same league to that of a fossil fuelled power station – to put this in perspective it would require thousands of the large 2 MW wind generators to match the capacity of the new 2000 MW CCGT Power Station in Pembrokeshire – the reason why thousands of 2 MW wind generators would be required is down to the fact that these generators have a load factor of only 29 per cent or less. Chapter Three of this book gives more detail.

Remember also, a fossil-fuelled powered station will give a continuous output until its fossil-fuel runs out whereas wind generators, by their very nature, are at the mercy of the wind. There are some misguided people who think that as soon as the wind blows, meaningful electricity will be produced and the harder the wind blows the greater the output! Sadly this is not the case as a wind generator requires a 'cut-on' wind speed of about 16 kilometres per hour (10 miles per hour) to start generating electricity. They reach peak power output at around 53 kilometres per hour (33 miles per hour) and at very high winds speeds, greater than 80 kilometres per hour (50 miles per hour) the wind generator must shut down, otherwise structural damage will occur.

Due to the high winds in Scotland during September, 2011, many wind farms were curtailed or stopped for parts of the day as follows: on the 11th Sept, 2011, 11 wind farms, 12th Sept, 2011, 9 wind farms, 13th Sept, 2011, 16 wind farms; these are undisputed government figures. Indeed, millions of pounds are still being paid out to the owners of these wind farms when they are not producing any power - heads they win, and tails dear reader, you lose. It cannot be reiterated and stressed enough, that under conditions of no wind there is no power generation, and at high wind speeds there is no power generation - think of that very cold, dark day in winter, with heavy frosts and no wind, just when you need electrical energy for boiling water, making soup, heating the house (central heating pumps and control systems require electricity to function), or providing lighting, only to find the power is not there - then you will truly curse the 'monstrous machine' by the name of wind turbine.

One of the strangest arguments put forward by proponents of wind driven electrical energy is the claimed significant saving in the production of carbon emissions. How can this be, when wind generators, for a guaranteed continuance of supply, require backup by fossil fuelled power stations when the wind fails. Do these people not know, or care, that none other than the German power company E.ON has stated quite publicly that wind generation will need at least 90 per cent backup by conventional power stations to guarantee security of supply – the Spanish company, Iberdrola has suggested 75 per cent. In fact it can be argued that if wind generation requires this backup why build wind farms in the first place – could it be just to fill the pockets of immoral and greedy people?

Fossil fuelled generators are designed to run continuously. If these power stations cannot meet this requirement, then fuel consumption and emissions of key air pollutants generally increase. A vehicle analogy helps explain the reason: a car that operates at a constant speed (e.g.55 miles per hour) will have better fuel efficiency, and emit less pollution per mile travelled, than one that is stuck in stop-and-go traffic. Now, as we all know, the wind by its very nature is stop-and-go, and the result being little, or no reductions in carbon emissions, by shifting wind to conventional generation.

The vacuous argument of the 'wind generator fanatic' is put into even more perspective when emissions from American industry and the industrialisation of China, India, Brazil and other countries are brought into the equation. Do the proponents of wind power not realise that during 2005 China set out on a seven-year programme to build over 500 new coal-fired power stations – just imagine what this will do for global CO_2 emissions, bearing in mind the UK contributes to less than 2 per cent of global carbon emissions – don't these fanatics know that China is now the world's largest polluter having recently overtaken the USA.

In considering the case for carbon emissions how many of these misguided wind generator supporters have given any thought to the CO_2 emissions of both military and civil aircraft - do they not realise that a typical jumbo jet yields something like 520,000 tonnes of CO_2 emissions per year! The Department for Trade and Industry (DTI), which during 2007 became the Department for Business, Enterprise and Regulatory Reforms (BERR) stated the 66 MW Fullabrook Down, North Devon wind scheme, would, when fully operational, save almost 65,000 tonnes of CO_2 emissions annually. But it would necessitate more than eight of the schemes similar to Fullabrook Down to compensate the annual emission of just one jumbo jet – in considering holiday flights, I wonder how many wind generator worshippers refuse to fly and employ the wind by sailing to their various holiday destinations?

The very hot weather of July, 2006 should have concentrated minds to the stupidity of large scale wind generation in the UK - the media were warning of electricity shortages by the National Grid as a result of increase in demand from business and homes using air-conditioning. Indeed, Piccadilly Square, London lost its power and the BBC Welsh Television News showed a picture of a wind generator with its blades absolutely motionless and obviously producing ZERO electrical power.

What a disaster there would be if we were heavily reliant on this source of electrical energy - it would not just be air-conditioning at risk, but fridges, freezers, lighting, television and everything else that is dependent on electricity. What will happen to the computerised systems of banks, business, our public services, the hospitals, schools, petrol stations et cetera when the power goes off? Remember also that home computers will shut down and deny access to the Internet – I hope you do not rely on the Internet for shopping, banking or anything else – it will prove more than troublesome if you do – and it is just not the summer months that would be perilous as the winter period would be far worse when there is little, no wind, or too strong a wind.

When I started researching and writing this book back during December 2006, the temperature outside was hovering around zero and there was no appreciable wind - when I ventured outside I could see three wind generators on the distant hills and not surprisingly, there was only one generator with its blades moving very slowly - thank goodness, I thought, that we do not have to rely on this source for electrical power. I really do shudder to think of what might be in store for all of us in the future if we continue with the current madness. With the advantage of hindsight, it could be claimed the power blackout (see Chapter Three for more information) over Europe during November 2006 was indeed the writing on the wall, but it appears nobody is taking any notice!

The power company E.ON, has admitted that without the hidden subsidy in our power bills nobody would be building wind farms. It seems incredible that we are faced with such totally blinkered and mindless policy making and it is time that everyone recognised the economic insanity it represents. The European green lobby is driving a compliant UK Government toward the brink of a precipice with a total disregard of the consequences for the people of the UK.

It is unbelievable that wind supporters, apart from the wind industry and landowners, continue to ignore the evidence and argue a case for the huge expansion of the wind generation both onshore and offshore. The Government should stop any further funding of wind generation technology and resolve, or ignore any issues that such action might create with Brussels. The Government, by its failure to make timely decisions regarding the replacement of electricity generation capacity, has already totally failed the public. Further pursuit of its absurd wind policy, which is distracting attention away from the hard decisions that need to be made as soon as possible, will be totally unforgivable.

It is shameful that Government has effectively masked other sustainable, renewable sources of energy by promoting the madness of large scale wind generation. There are other more worthy renewables such as tidal energy in the form of estuary and ocean impoundments, ocean current turbines and wave power, the use of micro-hydro technology in the UK. Not forgetting solar water heating and photovoltaic (PV) cells, geo-thermal and fuel cells...not to forget nuclear fusion, the power of the Sun.

Finally, why is the government so slow in developing shale gas in the UK – it has transformed the economy of the USA and would prove a salvation for the parlous state of the UK economy – currently we should be building more cheap, relatively clean and efficient gas-powered stations, whilst pursuing more environmentally but effective forms of generating electrical power – but make no mistake the global energy market is about to transform with the abundance of shale gas.

> "When the Paris Exhibition (of 1878) closed,
> electric light will close with it,
> and no more will be heard of it."

Oxford Professor Erasmus Wilson (1809 – 1884)

CHAPTER ONE

A Brief History of Electricity

The electrical energy we use every day is either in the form of Direct Current (DC) or Alternating Current (AC). We generally come across direct current in dry cells or batteries - when we purchase such items as electric torches, portable radios, and digital cameras the cells or batteries are usually provided - sometimes we may have to purchase the cell or battery separately. The voltage of these cells or batteries is of a low value such as 1.5 volt, 3 volt and 9 volt.

Larger 'wet type' of batteries are also available such as the 12 volt lead-acid version for cars, caravans et cetera; even larger lead-acid cells can be found in industry, such as in telephone exchanges, which have batteries supplying a total of 50 volts DC at the exchange bus bars.

Large modern power stations on the other hand produce alternating currents, and it is this electricity that is distributed across the country to industrial sites, offices and our houses.

In our homes we use alternating current to power such things as lighting circuits, electrical cookers, washing machines, electrical fires, television sets, home computers, radios, music centres and DVD players. Although it should be noted, that numerous items in the home such as televisions actually function on DC and at a much lower voltage level than the AC voltage fed by the power companies – thus components such as transformers and rectifiers have to be built into these devices to obtain the correct DC voltage levels.

With regard to exactly where our electricity comes from, it is interesting to ponder on how many people actually realise and appreciate that each household, every shop, office block and factory is connected via wires and cables to a large network of interconnected, synchronised power stations. Indeed, the whole of the UK electricity system is comprised of large alternating current power stations, interconnected by high voltage transmission lines, to carry electricity long distances ranging from tens to hundreds of kilometres. These high voltage transmission lines in turn feed distribution networks at a much lower voltage to supply electricity to consumers.

It is sobering to realise that the filament light bulb was invented nearly one hundred and fifty years ago during the 1880's, a national grid set up in the UK in 1926, centralisation and nationalisation established in 1946, the first nuclear power station built in 1956 at Calder Hall, with liberalisation of the power industry in 1990 followed by retail competition to customer level during 1998.

I doubt if many readers have ever thought about or enquired why a potentially fatal alternating voltage of 230 volts is provided at customer's premises - the United States inventor and scientist, Thomas Alva Edison (1847-1931) was reported as saying, "My personal desire would be to prohibit entirely the use of alternating currents. They are unnecessary as they are dangerous...I can therefore see no justification for the introduction of a system which has no element of performance and every element of danger to life and property." (Quoted from and acknowledgements to R.L. Weber, A Random Walk in Science).

According to the UK charity Electrical Safety First (www.electricalsafetyfirst.org.uk) accidental domestic fires of electrical origin during 2011-2012 accounted for 20,403 fires, 2469 injuries and 46 deaths. During 2011 a total of 2.5 million people, above 15 years of age, received mains voltage electrical shock, and of this total 350,000 sustained serious injury.

Electrical Safety First is dedicated to reducing deaths and injuries caused by electrical accidents, and aims to ensure everyone in the UK can use electricity safely. They campaign on behalf of consumers and electrical trade professionals to improve safety regulation and ensure safety messages are appropriate, up to date and well communicated. They provide expert information and advice to help protect people from faulty, damaged, sub-standard, and poorly maintained electrical installations and electrical appliances – being recognised by government and industry as the leading campaigning charity and technical authority on electrical safety.

So why a potentially dangerous alternating voltage of such magnitude you may well ask, and why not a direct voltage at a much lower level - especially when numerous devices in the home actually function on direct current.

The devices in a typical modern home that require direct current are typically television sets, home computers, music centres, radio's, DVD players and telephone answering machines - all these devices do not require an AC supply and need to operate on a direct voltage at a level considerably lower than 230 volts.

The above is the rear view of the system unit of a typical desk top computer. The mains power cable is seen at the top adjacent the power unit fan. Fundamentally the system unit requires DC with a maximum voltage of 12 volts.

Home computers (including the mains/battery operated lap-top computer) for example, need to transform the mains voltage down to much lower levels and then convert (rectify) the alternating voltage to that of a direct current, and a fan cooled, typical home desk top computer power unit would have the following specification:

AC Input: 115 Volt/230 Volt, 47 – 63 Hertz, 205 Watt.

DC Output: +5 Volt maximum current 20 Amp, +12 Volt maximum current 8 Amp, -5 Volt maximum current 0.5 Amp, -12 Volt maximum current 0.5 Amp.

We can see from the above specification that the maximum voltage the computer 'actually' needs is 12 volts DC at a power rating of 205 watts - so a supply voltage of 230 volts is, to say the least, way over the top. No doubt many readers have portable radios that will operate on a mains supply or batteries. Indeed, I have portable radio/cassette player that has the following specification:

AC Input: 100 – 110 V, 115 – 127 V, 200 – 250 V, 50/60 Hz, 6 W.
DC Input: 9 V. (Batteries: 6 x 'D' size).

The above specification, in effect, tells you the radio actually requires DC at 9 V and if plugged into a mains socket the AC mains has to be stepped down via a transformer and then rectified to this DC requirement.

Thus we can see that the mains voltage is far too excessive when we fully appreciate the actual level of, and type of voltage required for electronic devices. If we consider other requirements in the home such as lighting, we can argue that light bulbs and any other means of illumination can, and should operate at much lower voltage levels. Currently we have Energy Saving Bulbs (ESBs) that operate on mains voltage, but have a much significant lower power consumption than conventional bulbs for the same level of brightness; a typical example ESB specification being as follows:

Brightness: 20 W = 100 W.
Power consumption: 20 per cent (taking a standard 100 W light bulb as 100 per cent).
Lamp life: 10,000 hours.

But why not bulbs operating at say 12 volts DC - car lights operate at this level of voltage and can be extremely bright as we are all aware. Then consider Light Emitting Diodes (LED's) – a very real energy saver – these are discussed in more detail later in the book. Large power consuming items in the home such as electrical fires, stoves, and washing machines could all be designed (if the will is there) to operate effectively at a much lower level of voltage. So, why are homes still fed, in this day and age, with a 230 volt alternating supply and not a direct current at a much lower level?

To see how this has come about we need to explore the history behind the development of electricity as a source of energy in the home. It should be fully realised that when the early power stations were being developed some in fact did produce direct current, whilst others produced alternating current supplies. Taking Cardiff, the capital city of Wales, as an example, many years ago one of the main shopping areas, namely Queen Street, had a DC supply on one side of the street, whilst an AC supply on the other side of the street. A further example of direct current usage being the Royal Opera House, London which relied on a DC system operating from its basement up until 1995.

But I am getting ahead of myself and it is necessary to travel back in time to the advent of the electric telegraph to fully appreciate the transmission of electrical energy along a metallic conductor. Indeed, the first transmission of electrical energy along a wire was not for lighting or heating, but purely for the purpose of COMMUNICATION and it was when Sir Charles Wheatstone (1802-1875) and William Cooke (1806-1879) invented the electric telegraph in England, when about the same time Samuel Morse invented his system in the USA.

The Wheatstone and Cooke telegraph was patented back in 1837 and it was known as the five needle system, but the system was not very practicable being laborious and clumsy to use, needing five wires, which proved expensive to cable between transmitting and receiving ends. In this first commercial method of communication the indicator at the receiving end consisted of a vertical board on which the letters of the alphabet were marked in a diamond formation. An electric current was passed along a wire from the transmitting end to the indicator board at the receiving end, which had the alphabet marked out on it. Magnetic needles were suitably placed on the board and behind the needles were coils of wire; therefore when an electrical current passed through the coils, the needles were made to deflect so that two needles would point towards the required letter.

At the transmitting end a number of flexible metal strips could be depressed to make contact with metallic points below them. Thus the letters of the alphabet were so arranged in conjunction with these keys that the depression of two particular keys at once resulted in the magnetic needles at the receiving end to swing on their pivots in the required direction.

Cooke and Wheatstone had their first commercial success with a telegraph installed on the Great Western Railway (GWR) over the 13 miles (21 km) from Paddington station to West Drayton station in 1839 - this was the first commercial telegraph in the world.

The instruments used were Cooke's 'improved' four-needle telegraphs, which used a separate return wire so that signals could be made by converging two needles and by a single needle, making a total of twenty indications. The sixth wire was included as a 'spare'. The sending mechanism originally consisted of five of Cooke's rotating 'butterfly' commutators, although these were later replaced with Wheatstone's permutating buttons or keys.

Because of its limitations it was not surprising that the five wire system was soon replaced with a two wire system called the double needle system, which used a code to indicate the letters - the two-needle system was used with three of the remaining working underground wires, which despite using only two needles had a greater number of codes. But when the line was extended to Slough in 1843, a one-needle, two-wire system was installed - it was not long before the double needle method gave way to a single needle system which only required one wire and an earth return.

This system employed a special switch (commutator) at the transmitting end, which could reverse the direction of the transmitting current flow. At the receiving end a needle was arranged within a coil, which would deflect when a current was flowing through the coil. Thus depending on the current flow, as dictated by the transmitting end, the needle would swing one way or the other. The method lent itself to the Morse code as the needle swing in one direction can represent a dot and the needle swing in the opposite direction to indicate a dash. A single needle apparatus was patented in 1845 which was taken up by the railway companies at that time.

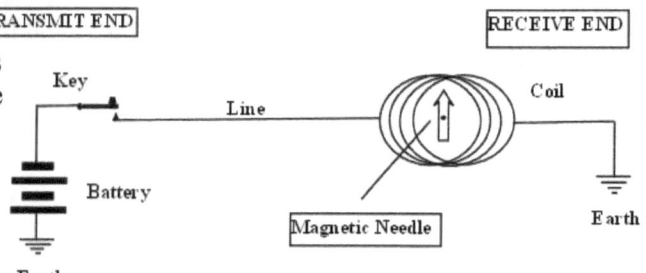

The above diagram shows a simple form of electric telegraph. When the key is operated at the transmitting end, a current will flow along the line wire and cause the magnetic needle to deflect within the coil at the receiving end. The above arrangement uses the ground as the return path, thereby completing the electrical circuit. All early telegraphs employed this method as it was obviously economical by not having to provide a return wire.

As mentioned above only one wire was used as the return path for the electrical current was via the ground, commonly known as an earth return. It should be clearly understood that at this time, the electrical energy used was direct current sourced from the chemical action of batteries. The system employing indicator needles was clumsy and laborious, but with the invention of Morse code in the United States during 1838, by Samuel Morse (1791-1872) and his assistant Alexander Bain (1810-1877), a more practicable system was achieved; this being based on the original alphabetical system in which pivoted and magnetised needles were still used.

Although an improvement on the original system it was still slow as it required the operator to watch the needle and write the message down at the same time; the system also operated as a single-line telegraphy or one-way traffic. When a sounder was produced by Samuel Morse the system was greatly improved as audible signals could now be received and a double contact key replaced the single contact key resulting in a considerable increase in operating speed. Additionally, a system was developed that allowed two-way messages on one line.

To improve the telegraph further an automatic system was introduced. This was achieved by utilising a machine very much like a typewriter, and the machine punched holes in a paper tape, so that the message is translated into Morse code on the paper in the form of perforations. This punched holed paper tape was then driven through a suitable transmitter by a toothed wheel. Speeds of 200 to 300 words per minute could be obtained with this method compared with 15 to 20 words per minute of the completely manual driven system.

The first Morse telegraph message was sent on 24 May 1844. The message was sent from Annapolis Junction (near Baltimore) to the capitol, Washington in the United States. Samuel Morse kindly allowed the daughter of a friend to choose the words of the message. The daughter, Annie Ellsworth, chose from a Book of the Holy Bible, namely, Numbers chapter 23, verse 23: 'What hath God wrought'.

The first printing telegraph to be used widely was developed during 1855-1863 but it did not use Morse code but a code of its own. Although the most successful system was developed by a Frenchman named Maurice Baudot (1845-1903). The system was employed extensively on British and continental lines; the complete Baudot system used a five unit code and was capable of transmitting eight messages simultaneously with speeds of up to 30 words per minute over one line.

During 1852 a submarine telegraph cable was laid across the English Channel enabling direct communications between London and Paris; by 1860 the telegraphs were working well and England was connected to the continent by submarine cables. It is amazing to think that Western Union in the United States have been sending telegrams for 150 years with the last being sent in the early part of 2006. The company was formed during April 1856 and built its first transcontinental telegraph line in 1861. Telegrams reached the top of their popularity during the 1920's and 1930's when they proved cheaper than long distance telephone calls; customers would use the word 'stop' instead of actual full-stops simply because punctuation was extra while the four-character word was free.

Next onto the scene came the telephone invented during 1872 by Alexander Graham Bell (1847-1922) of the United States. Initially all telephone circuits had a common earth return and as a result suffered from interference from telegraph services which also, as we have learnt, employed a common earth circuit. As these stray (earth) electrical currents became more and more serious all systems had to go to double line systems - this approach thankfully resolved the problem.

With the development of the electrical generator (Hippolyte Pixii, 1832), electric motor (Thomas Davenport, 1835) and the electric arc lamp (Pavel Yablochkov, 1875), electricity could now be used for other applications other than communications. We have seen that electricity was first used commercially for the electric telegraph which was powered initially by batteries - next onto the scene came the telephone, electrical generator and motor, electric arc lighting by which a generated current jumped a gap between carbon electrodes and produced a white-hot continuous spark (arc). It should be noted, that developments such as electric arc lighting utilised their own complete circuits as part of their design, which included the necessary batteries or generator to supply the electrical energy. Arc lighting initially was a system complete in itself when installed in a customer's premises. It included all the necessary lights, wiring and switches plus a stand-alone generator powered either by a steam engine or water wheel.

Unfortunately for arc lighting, apart from being extremely bright, it soon became apparent that the carbon electrodes burnt out very quickly and it was also noisy and smelly. Thus it was 'pioneers' such as Edison who had the initial vision of a central generating station with a network of wires feeding out to all of its customers - a system that would be superior to the existing network of gas pipes that fed both street and customer gas lights at the time. It is sobering to note that the first installations for gas lighting goes back to the 1790s supplying the stately homes in Britain; this being followed by the electric arc lighting systems. We have also seen that the telegraph and telephone systems were the first commercial network systems insomuch that they delivered (transmitted) electricity, albeit in the form of communication, over long distances to its customers; the telephone industry, by its very nature, having many more customers.

The world's first central electrical power station was built in the United States, by Charles Francis Brush (1849-1929). Charles Brush designed and developed an electric arc lighting system that was adopted throughout the United States and abroad during the 1880's. It should be noted the arc light preceded Edison's and Swan's incandescent light bulb in commercial use and was very suited to places where a bright light was required, such as street lighting and illuminating public and commercial buildings. A fundamental requirement in Brush's arc lighting system was the dynamo (electric generator). This generator was the workhorse of the centralised power station - a concept developed independently by Brush and Edison and which eventually evolved into the electrical power generating industry we know today. Therefore prior to Edison, Charles Brush had a centralised power station operating in New York City during 1881. Thomas Edison's centralised power station did not materialise until 1882 and was situated in Pearl Street, lower Manhattan, New York City. It started producing electricity during September of that year illuminating Wall Street offices including those of the New York Times.

These early power stations generated direct current (DC) and they were small and unable to distribute electrical energy over any appreciable distance due to the loss of energy in the line conductors in the form of heat. When electricity passes through a metallic line conductor it suffers from power loss obeying the law, Power (P) = Current squared (I^2) x Resistance (R); known commonly as the I^2R losses, or power losses. Therefore to transmit a useful current over a large distance would require metallic line conductors of large diameters which are very expensive, thus making direct current transmission over large distances uneconomical. Due to his marketing techniques, Edison soon became a dominant presence in the early central station electrical industry. Edison realised that doubling the output of a steam-powered generator did not double its capital cost, and a larger steam engine was more fuel-efficient, generating more electricity from a given amount of fuel. Although Edison could obtain a large steam engine, he had to design a generator that was compatible to it! He also had to construct cables sufficiently durable and effectively insulated to carry the currents over long distances - currents much greater than those used in telegraph or telephone cables.

Edison also realised that the many thousands of lamps operating off his network of wires could not be connected in series, for the very obvious reason that if one lamp was switched off or failed, then current would cease to flow in the network rendering all the other lamps useless. He realised they all had to be connected in parallel, which in itself presented a problem - with a low resistance element lamp, as more and more lights were switched on, then more and more current would flow in the network - this would demand very heavy (low resistance) and expensive cables, if the current were not to melt them. The way out of this dilemma was resolved when Edison and his staff, at their Menlo Park laboratory, during 1879, produced a carbonised bamboo fibre in an air-free globe - the incandescent bulb - this had a high resistance which presented Edison the breakthrough to push ahead with his large-scale central station electricity system.

It should be noted the invention of the incandescent bulb by Edison in America was challenged by Joseph Swan of Newcastle-upon-Tyne, England, who demonstrated a very similar design. Nevertheless the invention of the incandescent bulb at more or less at the same time, greatly hastened the demise of the electric arc light.

The early electric light bulbs used a filament of carbon drawn out to a very fine wire and enclosed in a bulb which had all the air extracted. The decision to use carbon was simply that it could be heated to a higher temperature, than any of the metals known at the time, and exclusion of the air prevented the filament burning away. Unfortunately, the temperature was limited and the carbon lamp gave off a very reddish light; during the early 1900's certain rare metals, having a high melting point, were used in an evacuated bulb.

Then in 1913 the Tungsten filament bulb was developed. This used a gas-filled bulb filled with an inert gas such as argon, into which was placed a Tungsten filament which could run at a temperature of about 3,5000C. This bulb emitted a light nearer to daylight than any of the previous lamps, although still containing too many red waves and not enough of the green and blue; additionally, the proportion of light waves to heat waves is greater in this type of lamp, so that more light is obtained for a given amount of power than in the earlier types. The amount of light emitted by a lamp is measured in 'lumens' and a small gas-filled lamp gives approximately 16 lumens of light for every watt of electrical power which it consumes. Therefore an ordinary 60-watt light bulb used for domestic lighting will give out about, 60 x 16 = 960 lumens of light.

The early central generating stations were in effect, analogous to a gas-making plant. The generating station fed electrical current through a 'loop' of wire extending for miles to customers premises, to which electrical lamps would be connected. These wires, as mentioned above, would be similar to telegraph cables, although being much more robust, because of the heavy currents they had to carry and maintain. Thus this 'electrical mains' and electrical lighting was analogous to the 'gas mains' and gas lighting at the time.

A major disadvantage with the early central generating stations is that they all produced direct current, that is, current that flowed in one direction only. Although the direction of current flow was not that critical, the problem was due to power loss in the cables - as the system expanded and the wires extended further and further, the losses from the wires became more and more expensive! The use of heavier and heavier wire was not the answer as they also became more and more expensive. So the direct current system had its limitations - but an answer was found in the use of alternating current electricity.

With the advent of the transformer (1831) it was found this new device could raise or lower the voltage of an alternating current as required, and we know that in an electric wire carrying a given amount of power, the higher the voltage, the lower the current and therefore the lower the losses in the wire. Therefore an AC generator at a power station could have its voltage output stepped up by a transformer - this stepped up voltage could then be transmitted over the network wires with minimal power losses - at the far end the voltage could then be stepped down, for customer usage.

Thus an AC system, due to its minimal power losses, could cover a substantially larger area than a DC system. It also meant that electricity could be generated at remote sites where water was freely available and could be employed to produce what is known as HYDROELECTRICITY and indeed, in the early days most electricity was produced this way and in many parts of the United States electricity and was commonly known as 'hydro'.

The provision of high-voltage electricity became known as TRANSMISSION so as to distinguish from the low-voltage electricity that was fed to customers, this low-voltage being called DISTRIBUTION. George Westinghouse of the United States was a leading pioneer in the use of AC and was one of Edison's most challenging rivals. During the 1880s there was fierce competition between the DC and AC systems, each having its various merits. The DC system was relatively simple and not so costly for customers close together in city centres. Whereas AC was more economical for customers spread over a large area in the suburbs and rural areas. Additionally, in those early days DC could run motors whereas AC could not. It was during the 1880's when 'three-phase' was introduced - which reduced the number and size of wires - that the AC system had another cost cutting boost. Thus with the advent of the transformer offering the ability to transmit electrical energy over large distances, *resulted in large central alternating current power stations becoming economically viable.*

The growing success of the AC system meant that hydroelectricity became very popular, especially in the United States, due to the fact that there were plenty of water resources. One of the most famous engineering projects in the United States was the construction of a large hydroelectric dam across the Colorado River, known as the Hoover (Boulder) Dam.

Countries such as the United Kingdom did not have the same water resources as the United States, but luckily they had plenty of fossil fuel in

USE OF TRANSFORMERS IN ELECTRICAL POWER DISTRIBUTION

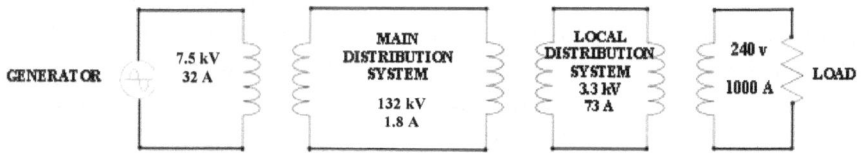

The power wasted as heat in the cables is proportional to the square of the current. Therefore the use of transformers enables the current in the long-distance cables to be of a much lower value than that drawn by the consumer's load. Thus a considerable saving results from the reduction of current in the main cable lengths. This results in much lighter cables being able to be used with a consequent saving in cost to both the cables and of the pylons, not forgetting the insulators needed to support them.

the form of coal. So it is not surprising that Britain's early power stations were fired by coal. Unfortunately the burning of coal covered the local neighbourhood with smoke and soot and then there was the problem of getting rid of the ash. Additionally, the noise and vibration of the early steam engines shook the surrounding buildings; living near one of these power stations was not an endearing experience. The early steam engines were not very efficient in extracting the chemical energy out of the coal and had 'fuel efficiencies' of less than ten per cent, that is, over ninety per cent of the energy in the coal was wasted! Indeed, the 'thermal efficiency' of steam locomotives and steam engines over a century ago was of the order of 1 to 2 per cent, whereas the 'thermal efficiency' of a steam turbine is about 40 per cent.

Thus it was when Charles Algernon Parsons (1854-1931) invented the 'Parsons steam turbine' for marine use during 1884 that improvements came in efficiency and smoother operation. Parson turbines were fitted to such ocean going liners as the Lusitania and Mauritania; his own steamship Turbinia reached a record breaking speed of 34.5 knots in 1897. Indeed, it was so successful that the Admiralty instructed Parsons to make a turbine-engine destroyer called H.M.S. Viper which was capable of 37 knots and faster than any previous naval ships in the world. Charles Parsons was born in London and set up his own company near Newcastle-upon-Tyne during 1889, where he developed turbo-generators of various kinds and increasing capacities. These formed the fundamental machinery for driving alternators in the production of electrical energy and were adopted on a national and international scale over a period of time.

As electricity became more and more popular numerous power stations were built, to the extent that nearly every town had its own power station with large towns and cities having more than one power station. The small town power stations generated about 4,000 to 5,000 kilowatts of power, whilst large places such as Manchester and Birmingham had bigger power stations capable of 50,000 to 100,000 kilowatts of power. It was soon recognised that electricity could be generated more economically in large power stations than in small ones. Additionally, every power station must have at least one and probably two spare generators, together with the corresponding turbine and boiler equipment, which can be used in case of a breakdown of one of the regular units - having spare generators meant that maintenance was easier to attend to as well.

It was also realised, that electrical energy could be generated much more cheaply by connecting all the towns and cities by transmission lines and actually generating all the electrical energy in a few large up-to-date stations, than by having numerous small and inefficient stations in each town and city. Using this arrangement the amount of spare plant could be reduced, as in the event of a breakdown at one power station then this electrical loss could be made up by another power station connected to the transmission lines.

Thus from 1917 to 1925 some steps were taken to link up a number of the towns in this way and some of the oldest power stations were closed down, although not much progress was made. However, a committee was set up in 1925 with Lord Weir as Chairman to examine the whole problem. As a result the committee suggested that certain of the largest and most efficient power stations should be selected and used for supplying the whole of the country with electricity whilst all the other power stations should be closed down. All the towns and cities should be interconnected by a network of high voltage transmission lines criss-crossing the country in the form of a large grid network. The committee suggested that by employing such a scheme the whole of the country could be supplied with cheaper electricity.

An electrical grid in the United Kingdom was set up during 1926 followed by large scale 'interconnectors' during the 1930's - the purpose of interconnection was to produce electricity in the most efficient and economical way. To achieve this aim generators had to be large and kept as fully loaded as possible at all times, the electrical industry terminology being: to achieve a 'high load factor'. This was brought about by generating electricity in a few large power stations, rather than running a number of small and independent plants. Also by the utilisation of high-voltage lines, known as a Grid, major plants were interconnected so that they could operate together. As mentioned earlier the pooling together of resources of a large area had the effect of lessening the capacity for spare plant, significantly offering a reduction in capital costs. Since the system employed alternating current, synchronisation and voltage level control was essential.

All generators at power stations are connected in parallel to the main station busbars. In the case of the direct current power stations there is not a problem in running up and connecting a standby generator to the main bus bars when the network load demands it. As soon as the generator reaches operational speed and the required voltage level, the generator was simply connected to the main busbars by means of adequate switches.

But in the case of alternating current power stations this cannot be done due to the nature of the alternating current - if it were connected in the same manner as a direct current generator then considerable damage would be done, possibly resulting in an explosion of one or more of the generators - thus before any such connection certain criteria must be satisfied:

- ❖ The voltage of the generator to be connected must be the same as the station main busbars.

- ❖ The voltage of the generator to be connected must be in phase with the station main busbar voltage.

❖ The frequency of the generator to be connected must be the same as the station main busbar frequency.

It must be remembered that with alternating current generators their voltages are varying from instant to instant, so it is critical that some means must be found to ensure the connecting generator is the same voltage and frequency at the instant of connecting at the main busbar.

Thus without going into too much detail, as it is beyond the scope of this book, this condition could be achieved by the use of lamps: The 'Lamp Dark' or the 'Lamp Bright' method.

The 'Lamp Dark' method necessitates two lamps, each of the same voltage as the busbar voltage. Each lamp is connected across the poles of the standby generator connection switch. One lamp is connected to the main busbar and one pole of the standby generator connection switch and another lamp connected to the main busbar and one pole of the other standby generator connection switch. Remember although the standby generator has one switch there are two poles to this switch, which connect it to the power station busbars. Thus when the standby generator is running and generating the required level of electromotive force (emf) the lamps will remain consistently dark. This means there is no voltage difference between the machine and the busbar voltages and the switch can safely be closed as all the required conditions are met.

In the 'Lamp Bright' method the connections are different such that the 'switching in' of the standby generator can only be made when both lamps glow at their maximum brilliancy.

For large turbo-generators the operation of synchronising must be carried out very accurately and a special instrument called a 'Rotary Synchroscope' is employed. The synchroscope fundamentally consists of a small motor, housed within an instrument case, with the motor shaft connected to a pointer needle. The stator winding of this alternating current motor is connected to the power station main busbars and the rotor to the terminals of the incoming (standby) generator. A difference in frequency causes the pointer needle to rotate in a clockwise direction if the speed of the in-coming generator is too high and in an anti-clockwise direction if the speed is too low. When the two frequencies are identical, the pointer needle remains stationary at a position proportional to the phase difference between the two voltages.

Thus when synchronisation is reached, the pointer needle stands vertically upright. It should be noted that the connections to the coils of the alternating current motor are made through step-down voltage transformers when the busbar voltage is high.

From 1947 the majority of large power stations and the Grid in the United Kingdom came under the ownership of the Central Electricity Generating Board (CEGB). The large power stations were interconnected by the Board's 132,000 volt grid lines.

By 1957 the integration and nationalisation of the electricity system in the UK comprised of the Central Generating Board, 12 distribution boards and an Electricity Council for England and Wales. The north of Scotland, the south of Scotland and Northern Ireland although separate organisations, were still owned by the government. Today we have a privatised network after liberalisation in 1990, and all of the huge generating stations have almost completely replaced the smaller stations.

Modern power stations fall into three main categories:

1. Fossil fuelled (Coal, gas and oil).
2. Nuclear.
3. Hydroelectric.

All these stations employ large alternating voltage generators (alternators) to produce electrical energy and have the generator voltage 'stepped up' for line transmission, being reduced to a suitable level at the distribution end.

The amount of electricity actually used in a year, as a fraction of the maximum amount the system could supply, came to be called the 'load factor', a concept first introduced in 1891. The equation for load factor = Actual amount of power produced over time/ Power that would have been produced if generator operated at maximum output 100 per cent of the time. But in considering the equation it is important to fully understand some of the variables that govern the numerator, which we shall call variables 1, 2, 3 and 4, see below:

1. Deliberate operator intervention of generator/station output as a result of the economics of running the generator/station at any one time i.e. generator/station only run when price is right.

2. Necessary scheduled maintenance of generator/station.

3. Uncontrolled restriction of generator output due to lack of generator driving force such as "No Wind" in the case of a wind generator or severe summer drought as in the case of a hydro scheme.

4. Downtime due to unforeseen generator/station faults.

Now apart from variable 1, which is owner controlled, load factor identifies the overall effectiveness of a generator/station output and is therefore useful in making comparisons between different types of fuelled generators/stations such as coal, gas, oil hydro, nuclear and wind in the production of electrical energy.

Thus a generator/station with a load fact of 0.6 is far more effective than a generator/station with a load factor of 0.25 bearing in mind the influence of variable 1. It should be noted that Load factors are generally quoted as a percentage e.g. 60 per cent or 25 per cent.

Currently the National Grid's transmission network in England and Wales consists in the vast majority of 275 kV and 400 kV transmission lines, although there are about half a dozen short lines at 132 kV that are classified as transmission; most 132 kV count as distribution lines. In the Scottish Power area (i.e. the South of Scotland) transmission are defined as 132 kV and above. In the Scottish and Southern area (i.e. the North of Scotland), transmission lines count as 132 kV and above, but also includes some 33 kV due to a mixture of history and Scottish demography.

Summary

We have seen that the first transmission of electrical energy over a distance was for the purpose of communication, starting with the electric telegraph through to the conveyance of voice transmission with the advent of the telephone. The first electrical generators produced direct current for electric arc lighting. After the invention of the incandescent bulb came the rapid growth of the direct current centralised power station. Although it soon became apparent that to reach a large number of customers spread over a large area direct current had its limitations due to power loss in the network cables. To feed direct current over any appreciable distance would require conductors of large cross-sectional area and this proved prohibitively expensive.

The invention of the transformer overcame the problem if alternating current were to be used as it enabled the voltage to be stepped up with a significant decrease in the current. It must be remembered that when current flows through a conductor (wire) it encounters resistance which makes the conductor heat up, therefore energy is lost in this process before it reaches the user. It is important to note that for a given conductor (wire) the doubling of the current quadruples the losses according to the law I^2R (where I = the current, and R = resistance) and commonly known as the power loss.

For a while there was a battle between DC and AC systems as apart from the line losses direct current generators were easily brought into service at power stations, whilst alternating current generators required voltage and frequency compatibility before they could be brought on line.

Due to the superiority of alternating current for long line transmission, the synchronised alternating current system soon became the standard, as large power stations could be built, for example, where there were ample supplies of water for hydro systems.

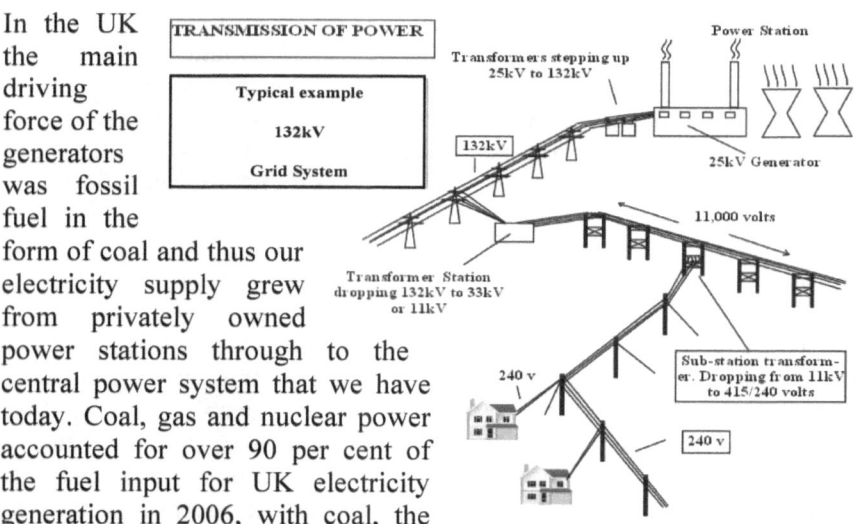

In the UK the main driving force of the generators was fossil fuel in the form of coal and thus our electricity supply grew from privately owned power stations through to the central power system that we have today. Coal, gas and nuclear power accounted for over 90 per cent of the fuel input for UK electricity generation in 2006, with coal, the largest fuel group, accounting for 41 per cent alone. Since 1976 the total UK electricity used has increased by 58 per cent, although there has only been a 30 per cent increase in fuel input over this period due to greater efficiency in electricity generation.

It is interesting that household domestic appliances consumed double the electricity in 2005 as they did in 1976 – but this should not come as a surprise when consideration is given to the current profusion of electrical items now found in the home – we have become so dependent on electricity that cuts in electrical power would have far reaching effects these days – indeed, a total loss of electrical power would prove catastrophic for our highly technological civilisation.

Finally, the reader should be aware of mains fluctuations and possible power failures - if a multi-meter is carefully (remember mains voltages can be lethal) connected to the live and neutral of the house supply, it will be noted, depending on where you live, that the incoming voltage can be anything from between 216V to 253V, and the voltage can change by two or more volts in a minute.

Thus in an effort to minimise stress to sensitive electronic components and achieve better efficiency it is recommended that a voltage regulator is fitted to household supplies, see Chapter Seven on 'How to Reduce your Domestic Energy Bill' for further detail.

In times of power failure the voltage can drop down into the mid-hundreds, with most equipment struggling to operate effectively at this level and this is known as a 'brownout' - when there is a complete failure of the mains then this is known as a 'blackout'.

"It ain't what you don't know that gets you into trouble. It's what you know for sure that just ain't so."

Samuel Langhorne Clemens (1835-1910)
(Mark Twain)

CHAPTER TWO

Wind, Weather & Climate Change

Historians will no doubt eventually write about the great scam of large scale wind generation in the UK, and the preposterous claim that if we surround our shores and desecrate our beautiful landscape with gargantuan wind generators, then they will have an impact on global emissions – which, of course, is utter nonsense. Thus this book would not be complete without a chapter dealing with wind, weather and climate change.

But before doing so, it will be useful to remind ourselves of some basic physics we may have forgotten since leaving school, which will be instrumental in understanding the topics covered in this, and the following chapter.

We all realise that matter is composed of continually 'jiggling' atoms or molecules - the rate of the molecular vibrations determine the nature of the matter on whether it takes the form of a solid, liquid, gas, or plasma – by virtue of this vibratory motion, the molecules or atoms in matter possess kinetic energy.

Now the average kinetic energy of the individual particles is directly related to a property we can sense – how hot something is – which brings us to heat, or if you like, thermal energy.

A rise in the temperature of matter makes the particles vibrate faster, and the hotter the substance, the more its molecules vibrate and therefore the higher is the thermal energy (kinetic energy) - putting a saucepan of cold water on a heated gas ring will cause the molecules in the cold water to vibrate more and cause a rise in temperature of the water - after a period of time the water in the saucepan will reach boiling point. Strike a metal coin with a hammer and it becomes warm because the hammer's blow will cause the molecules in the metal to jostle faster. Pumping up a bicycle tyre will cause the air to warm due to the pump rapidly increasing the air pressure (pushing the molecules closer together), and the tyre cools when we let the air out, decreasing air pressure (the molecules move apart) thus giving a rise and then a fall in temperature.

So what is temperature? The temperature of an object is simply how hot or cold the object is, and is measured in degrees Celsius (^0C) after the Swedish astronomer Anders Celsius (1701-1744). It should be noted though that thermometers divided into a hundred divisions between the freezing and boiling point of water were called Centigrade thermometers - from *centi*, 'hundreth' and *gradus*, 'degree' – but now are called a Celsius thermometer in honour of Anders Celsius who first suggested the scale.

Thus a typical device for measuring temperature is called a thermometer. Nearly all materials expand when their temperature is raised and contract when it is lowered. A thermometer is a common instrument that measures temperature by means of the expansion and contraction of a liquid, usually mercury or coloured alcohol. These thermometers use a scale which is usually dived into a hundred divisions, with each division known as a degrees, with the number 0 assigned to the temperature at which water freezes (0 degrees Celsius), and the number 100 to the temperature at which water boils at standard atmospheric pressure (100 degrees Celsius). Temperatures can also be measured in the Fahrenheit scale, named after the German physicist Daniel Gabriel Fahrenheit (1686-1736) and is denoted by the symbol '^0F.' It should be noted that on the Fahrenheit scale, water freezes at 32 ^0F, and boils at 212 ^0F.

There is another temperature scale which is favoured by scientists and is known as the Kelvin scale, named after the British physicist Lord Kelvin (1824-1907) - this scale is not calibrated in terms of the freezing and boiling points of water, but in terms of energy itself. Thus the number 0 is assigned to the lowest possible temperature known as Absolute Zero, at which a substance has absolutely no kinetic energy to give up - it can be approached but not reached in any actual physical system. Absolute zero corresponds to minus 273 degrees on the Celsius scale, and minus 459.67 degrees on the Fahrenheit scale. Units on the Kelvin scale are the same size as degrees on the Celsius scale, so the temperature at which ice melts is 273 ^0Kelvin - noting that there are no negative numbers on the Kelvin scale.

Having reminded ourselves of exactly what is meant by heat and temperature we will now consider what is meant by the transfer of heat. If several objects are in contact with one another, then the warmer object will become cooler and those that are cool will become warmer – heat always transfers from warmer to cooler things – and they tend to reach a common temperature. This equalising of temperature occurs in three ways, namely, conduction, convection and radiation.

Conduction

Conduction is the transfer and distribution of heat energy that moves from molecule to molecule within a substance. Not all substances conduct heat at the same speed. Metals (such as copper) and stone are considered good conductors since they can speedily transfer heat, but wood, paper, air, and cloth are poor heat conductors. Poor conductors of heat are called insulators. Heat energy is conducted from the hot end of an object to the cold end.

Convection

Liquids and gases are fluids. The particles in these fluids can move from place to place. Convection occurs when particles with a lot of heat energy in a liquid or gas move and take the place of particles with less heat energy. Heat energy is transferred from hot places to cooler places by convection. Thus convection is simply the transfer of heat energy in a gas or liquid by means of currents in the heated fluid – the fluid moves, carrying heat energy with it.

Liquids and gases expand when they are heated. This is because the particles in liquids and gases move faster when they are heated than they do when they are cold. As a result, the particles take up more volume. This is because the gap between particles widens, while the particles themselves stay the same size. The liquid or gas in hot areas is less dense than the liquid or gas in cold areas, so it rises into the cold areas. The denser cold liquid or gas falls into the warm areas. In this way, convection currents that transfer heat from place to place are set up.

Radiation

We all know that heat from the Sun passes through the vacuum of space, then through our atmosphere and warms the surface of the Earth. It should be noted this heat does not pass through the atmosphere by conduction, for air (gas) is a poor conductor – nor does it pass through by convection, for convection begins only after the Earth is warmed. We also know that neither conduction nor convection is possible in the vacuum of space between our atmosphere and the Sun. Radiation energy is the energy of electromagnetic waves - there are many different types of radiation, indeed an entire spectrum of radiation, and the spectrum is known as the Electromagnetic spectrum. There are many different kinds of electromagnetic waves with all of them having different wavelengths, properties, frequencies and power, and all interact with matter differently – this spectrum ranges from the longest wavelengths, but lowest frequencies, which are radio waves, to the shortest wavelengths, but highest frequencies, known as Gamma waves. Between radio waves and Gamma wave we have the other parts of the spectrum - increasing in frequency and moving across the spectrum from radio waves we have microwaves, infrared waves, visible light, ultraviolet waves and x-rays – it should be noted that the shorter the wavelength, the higher its frequency and vice versa. Radiation is the transfer of energy at the speed of light at 300,000 kilometres per second (186,000 miles per second) by means of electromagnetic waves.

Wind

Wind is simply the movement of the gas molecules that make up our atmosphere. The atmosphere extends to 2,414 kilometres (1,500 miles) above the surface of our planet, with about 75 per cent of the gasses residing within 16 kilometres (10 miles) of the surface; the comfortable breathable part of the atmosphere being well under 8 kilometres (5 miles) from the surface.

The gasses which make up our atmosphere are as follows:

Nitrogen at 76.08 per cent.
Oxygen at 20.95 per cent.
Argon at 0.93 per cent.
Carbon dioxide at 0.032 per cent.

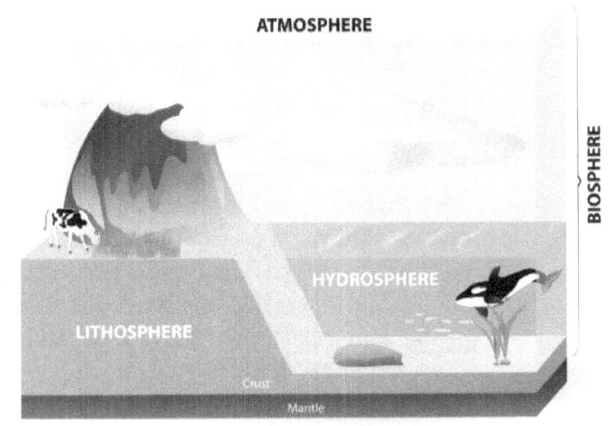

With the remainder of the atmosphere consisting of traces of Neon, Helium, Methane, Krypton, Hydrogen, Nitrous oxide, Carbon monoxide, xenon, water vapour and dust. The atmosphere is divided into several layers and working from the surface up, we have the Troposphere, which extends to 11 kilometres (7 miles) above the Earth's surface. Next is the Stratosphere, which contains the Ozone layer, and extends to 48 kilometres (30 miles) above the surface, then we have the Mesosphere extending to 88 kilometres (55 miles), then the Thermosphere extending to 700 kilometres (435 miles), and lastly the Exosphere which is an ill-defined region fading off into the vacuum of space. Obviously the part of the atmosphere this book is concerned about is the Troposphere, especially the lower part, since that is where all the action takes place as they say - it is where all living organisms (apart from marine and those beneath the ground) exist, and where the 'our weather' occurs, indeed, where the wind blows.

It can be argued that, fundamentally, wind is a converted form of solar energy - the Sun's radiation heats various parts of the globe at different rates with the greatest difference being between the Sunlit surfaces and those out of Sunlight (night time surfaces); dissimilar surfaces will absorb and reflect at differing rates, such as land, water and snow. These differences will have the effect of causing various portions of the atmosphere to warm up at different rates. Thus the warm air will rise reducing the atmospheric pressure at the Earth's surface, enabling cooler air to be drawn in to replace it - this movement of the air is what we know as wind.

But before discussing the wind further, and in an attempt to understand the nature of the atmosphere, it is useful to consider the atmosphere of other planets and in particular our two nearest planetary neighbours, namely, Mars being 34 million miles from the Earth at its nearest point, and Venus at 67 million miles distant at its nearest point – recognising that planets have elliptical orbits around the Sun.

Scientists claim that back in the mists of time Mars was a very different world to what it is today, for Martian geology offers evidence of dried-up river valleys, flood plains, sediment deposits, erosion features and evidence of coastlines showing that Mars, at some time in the past, was wet and warm. But over the past four billion years the Martian atmosphere has been escaping into space, leaving a very unwelcome place with a rarefied atmosphere and surface pressure just 0.6 per cent that of sea level on the Earth. The average temperature on Mars is about minus -55 degrees Celsius. Surface temperatures may reach a high of about 20 degrees Celsius at noon at the equator, and a low of about -153 degrees Celsius at the poles. The warmest soil temperature on the Martian surface estimated by the Viking Orbiter was 27 degrees Celsius, whereas the Spirit rover recorded a maximum daytime air temperature in the shade of 35 degrees Celsius, and regularly recorded temperatures well above zero degrees Celsius, except in winter. Water and carbon dioxide remain locked into sub-surface ice. So what has happened on Mars?

According to scientists Mars lost most of its atmosphere due to the core of the planet cooling down, and thus losing its protective magnetosphere, which in turn allowed the Solar wind to strip away much of the Martian atmosphere. The Earth, apart from retaining its magnetosphere, is much more massive than Mars - indeed it is virtually a thousand times greater with a mass of 6×10^{24} kilograms compared to 6.4×10^{21} kilograms of Mars. This greater mass of the Earth means that our planet has a greater gravitational grip on atmospheric atoms - it also means that the Earth is able to retain its internal heat for much longer, and the molten spinning core generates the Earth's magnetic field that protects us from the ravages of the Solar wind on a constant basis.

Another interesting point regarding Mars cooling down is that there are no plate tectonics - latest thinking suggests that plate tectonics is crucial for life on Earth. It is suggested that plate tectonics act as a geological thermostat that has stabilised our surface temperature over millions of years by regulating the concentration of carbon dioxide in our atmosphere.

As a small digression and regarding the possibility of life existing on a celestial body other than the Earth, where plate tectonics are active, then there is only one other candidate in the Solar System, and that happens to be one of the Galilean moons of Jupiter, namely, Europa. Scientists believe that deep below the ice on the surface of Europa lies a deep ocean – this liquid ocean holding double the amount of water as all the Earth's oceans, being kept warm predominantly by tidal heating as a consequence to the close proximity of Jupiter, and to a lesser degree, by natural radiation from the silicate rocks - the Galileo mission found strong evidence that a subsurface ocean of salty water is in contact with a rocky seafloor. Europa is slightly smaller than Earth's moon, and orbits Jupiter every 3.5 days and is tidally locked - just like the Earth's Moon - so that the same side of Europa faces Jupiter at all times. Europa is thought to have an iron core, a rocky mantle and a surface ocean of salty water, like Earth - this ocean though is deep enough to cover the whole surface of Europa, and being far from the Sun, the ocean surface is globally frozen over. Scientists are eager to learn if the reddish-brown fractures, and other markings spattered across the icy surface of the moon, contain clues about the geological history of Europa and the chemistry of the global ocean that is thought to exist beneath the ice - the cycling of material (due to tectonic activity) between the ocean and ice shell could potentially provide sources of chemical energy that could sustain simple life forms.

Due to the fact that Mars has lost most of its atmosphere it does not take a particularly large meteor to penetrate the Martian atmosphere and impact the Martian surface. The Context Camera on-board NASA's Mars Reconnaissance Orbiter spacecraft takes 30-kilometre wide black and white images of the Martian surface on a regular basis, looking for interesting surface images for the spacecraft's high resolution camera to further investigate. Interestingly between July, 2010 and May, 2012, the Context Camera images suggested that something had changed on the surface. Following up with the high resolution camera a 30-metre wide crater was noted with fresh dust and debris radiating out in dark spokes as far as 15 kilometres – statistics indicate that at least 200 new craters at least four metres in diameter appear on Mars every year – so it would appear that we have a lot to thank our thicker atmosphere for, or it could be tin hats all around!

A further example of how the thin Martian atmosphere reacts to celestial objects was the close encounter of Mars with a comet named C/2103 A1 on 19th October 2014. The comet C/2013 A1 (Siding Spring) is an Oort cloud comet discovered on 3 January 2013 at Siding Spring Observatory, Australia using the 0.5 metre (20 inch) Uppsala Southern Schmidt Telescope. Initially some observers thought the comet might have collided with the Martian surface, but in the event the comet passed within 139,500 kilometres of Mars with the comet dust ionising the Martian atmosphere – according to the Mars Reconnaissance Orbiter the comet had a nucleus of two kilometres across. The fine dust from the comet penetrated the Martian atmosphere resulting in a meteor storm of thousands of meteors per hour. To avoid being damaged by the comet's dust all the spacecraft orbiting Mars were moved to the far side of the planet for 20 minutes when the comet dust was at its most intense – the five spacecraft, namely, NASA's MAVEN, Mars Reconnaissance Orbiter and Mars Odyssey, Europe's Mars Express and India's Mars Orbiter, all came through the close encounter unharmed.

At the other end of the scale, so to speak, we have the planet Venus, known as Earth's twin, due to its similar size. This planet has a very thick, opaque atmosphere that is 98 per cent carbon dioxide, with clouds consisting of sulphuric acid and sulphur. This has resulted in a runaway Greenhouse Effect with temperatures as high as 470 degrees Celsius. This is hot enough to melt lead, and Venus has pressures 90 times higher than on Earth, as the early Russian space probes soon discovered to their cost when they landed there. Intriguingly Venus does not have a magnetic field and is closer to the Sun where the Solar wind is stronger. The planet has less water in its atmosphere than even Mars, this water possibly broken up by Solar ultraviolet light into its component parts of oxygen and hydrogen, and then lost in space. It appears that due to the mass of Venus this planet has been able to hang on to the rest of its atmosphere.

Coming back down to earth as they say, we will now consider the power in the wind. Since the atmosphere is made up of gas molecules it will have mass, and when this mass is in motion it will contain the energy of that motion; this energy is known as kinetic energy. We can see the result of this force in the movement of the clouds above our heads and the lashing of rain during a storm. We can feel the force of this energy when the wind blows in our face, and at times of gale force winds, we can be blown off our feet, and indeed witness the terrifying destructiveness of storm class winds - it is this energy that gives wind its ability to do work, be it constructive or indeed destructive as in the case of gales and storm conditions, with the extreme examples being tornados, hurricanes and typhoons.

In the past this kinetic wind energy was put to constructive use in 'pushing' along Roman Galleys, Viking ships and Galleons; today it is utilised to power various designs of sailing ships. The wind was also employed in moving the sails or vanes of windmills to grind corn et cetera; in many rural areas of the USA the 'farm windmill' (seen in many a western film) is commonly used for pumping water. The next chapter of this book examines how electrical energy is obtained from the wind by employing a large propeller to capture some of the energy in the wind to drive an electrical generator or alternator. These wind driven machines are commonly and incorrectly referred to as wind turbines, and the next chapter of this book will explain why this is so.

Fundamentally, when the wind strikes an object, it exerts a force on the object trying to move it out of the way - in doing so some of the wind's kinetic energy is transferred to the object. Depending on the strength of the wind and the nature of the object the wind is striking, there will be motion of the object such as paper or leaves moving across the ground, trees swaying or indeed, trees being uprooted and structural damage to properties.

The amount of energy in the wind is a function of its velocity (speed) and mass. Therefore the greater the velocity or speed of the wind the greater the energy, and the greater the mass of the air (wind) the greater the energy. The following equation gives the relationship between velocity (v), mass (m) and energy of the wind:

Kinetic Energy = ½ mv²

Now the mass of the wind is the product of its density and its volume; as the air is constantly in motion the volume can be determined by multiplying the wind's speed by the area through which it passes during a specific period of time. Therefore the mass of the wind can be represented by the following:

m = davt

Where,

d = density of air
a = area through which the wind passes
v = velocity (speed) of the wind
t = the period of time

The density of the air is dependent on temperature and altitude, which means the air, is less dense in summer than in the winter; it therefore follows that air density is greater at the poles than at the equator; increasing altitude also has an effect by decreasing the air density. Therefore any given wind generator will produce less electrical energy in the summer than it will in mid-winter with winds of the same speed – and will produce more electrical energy at sea level than say, 1500 metres (5000 ft) above sea level.

We have seen that the energy in the wind is given by:

Kinetic Energy = ½ mv²

If we substitute m in this equation we have:

Wind Energy = ½ (davt) v2 = ½ datv³

It should be noted that because water is 800 times more dense than air, coupled with the fact that water can be stored in large reservoirs, hydro-electrical schemes are far more effective and superior to wind driven schemes employing wind generators.

Power Available for Wind Generation

Now power is the rate at which energy is available – or the rate at which energy passes through an area per unit of time, therefore power:

$P = \frac{1}{2} dav^3$ watts

Therefore if the air density or the area intercepting the wind is increased then the power is increased - but it should be noted from the above equation the power in the wind varies with the cube of wind speed - now what exactly does this mean? In simple language it means that if the speed of the wind, for example, is doubled, then the power is increased by eight times. Thus a small increase in wind speed substantially boosts the power in the wind – this is why wind generators should be placed where the winds are best.

Previously, we have seen that the power in the wind is exponentially related to wind speed, insomuch that if the wind speed is doubled, then the power is increased by eight times. But power is also directly related to the area intercepting the wind – in the case of a wind generator, where the propeller rotates about a horizontal axis, it is the area which is swept by the propeller. This area takes the form of a disc whose area (a) can be determined by the equation: $a = \pi r^2$ where r is the radius of the propeller (or very nearly the length of one blade). Now relatively small increases in blade length (and therefore propeller diameter) produce a corresponding large increase in swept area, and therefore in power. For example, if we have a wind generator with a propeller spanning 2.4 metres in diameter subjected to a wind speed of 5.4 metres a second, and a power produced of 434.6 watts, then if we increase the propeller size to 3.4 metres for the same wind speed, we will increase the power to 868 watts – increasing the propeller size to 3.4 metres effectively doubles the area swept and therefore, as a result, doubles the power.

It may not be immediately obvious, but nothing tells you more about a wind generators potential than the size (diameter) of the propeller. Indeed, the wind generator with the largest propeller will almost invariably generate more electricity than a wind generator with a smaller propeller, regardless of generator or alternator size.

It is important to note that in propeller design there is a maximum in which they can capture the wind and this is the Betz Limit, named after the German scientist, Albert Betz (1885-1968) – he theorised that a portion of the wind must keep moving through the propeller, and not all of it could be captured – the Betz Limit is 59.3 per cent of the energy available to the propeller. Although experience has shown that wind generator propellers convert much less than this limit, with optimum designed propellers reaching levels slightly above 40 per cent Unfortunately, the usable energy from the wind is even less as energy is lost in the transmission machinery to the alternator and in the alternator itself - and unlike an anemometer (an instrument for measuring wind speed and which can measure gusts of wind) a wind generator does not respond to all gusts because of its inherent inertia.

Thus wind generators miss some of the wind, or more precisely the wind energy in gusts. Although this wind energy will be registered by an anemometer the wind generator may not 'see' the gust, especially if the speed of the gust exceeds the wind generator's operating limits. It should also be recognised that changing (Yawing) the direction of a wind generator as the wind changes direction also causes a problem – cup anemometers capture the wind from all directions, but a wind generator takes time to change its position and therefore misses a portion of the wind that is recorded by the anemometer. When all these loss factors are taken into consideration a good designed wind generator performing at 60 per cent of the Betz Limit can only deliver about 30 per cent of the overall energy available! Experience has shown that wind generators can only capture 12 per cent to 30 per cent of the energy in the wind.

A way of measuring wind force (at sea) was developed by Admiral Sir Francis Beaufort (1774-1857) in 1806. He developed the force scale (known as the Beaufort scale) based on the effects of wind on a canvas of a full-rigged frigate. The actual wind speeds were not measured until 1946 whereby the speeds were determined by use of the anemometer. The land scale is measured in miles (kilometres) per hour, whereas the sea scale is measured in knots.

Weather in the United Kingdom

Unlike most parts of the world the United Kingdom experiences very changeable weather. For example, if we were to compare the weather on two parts of the globe that share the same latitude, such as Winnipeg and the Lizard in Cornwall (both being approximately 50 degrees north of the equator), we should expect to find the same weather patterns, as they are both similar distances from the equator and thus receive similar amounts of solar radiation - but we find this is not so and there is a very marked difference in the weather.

During the early winter in Winnipeg it may be sunny with temperatures rising to about minus 12 degrees Celsius, but the air is sparkling and dry. By mid-winter, though, the maximum temperature may reach minus 29 degrees Celsius and if a wind is blowing very few people will venture outside unless it is necessary. The snow is on the ground for at least four months and often in the spring, the weather may be very blustery with a lot of snow storms rushing across the prairies.

How different to the not so cold, but wet and possibly stormy winters of the Cornish peninsular. Indeed, just 30 miles off the tip of Cornwall, the Scilly Isles can experience temperatures in the region of 10 degrees Celsius during January. Then in the summer most days will be sunny in Winnipeg with temperatures reaching 26 to 30 degrees Celsius, with the air being quite dry - a housewife may hang out her washing and it will be dry in 20 to 30 minutes due to the dry breeze – what a contrast to the summers experienced in Cornwall - so why should there be such a difference in the two weather patterns? One of the most important factors is with respect to areas of land and sea.

A land mass will gain or lose heat much more rapidly than a body of water - a large continent becomes cold in winter, thus cooling the air above it - when air temperature is lowered, its density increases and it will tend to sink. The pressure of the atmosphere, which we can think of as weight, will therefore be increased. Winnipeg is right in the middle of such a land mass and as a consequence its winter weather will be very cold. The subsiding air is also very dry, so there is little cloud and precipitation in the winter. It is interesting to note that the snow fall is only about 50 centimetres (20 inches) compared to 250 centimetres (100 inches) in the east of the continent, where more moisture is available from the ocean. In summer the land mass gains heat very quickly and during the summer months in Winnipeg the weather is warm and sunny, with the air being quite dry because of its distance from the sea.

An important 'force of nature' which influences our climate is the Gulf Stream and is the most important ocean-current system in the northern hemisphere. The Gulf Stream flows from Florida to north-western Europe, and incorporates several currents: the Florida current, the Gulf Stream itself, and after passing the Grand Banks (off Newfoundland) the flow forms the shallow, broad slow-moving North Atlantic Drift. This ocean-current system is a good example of how warm waters at the Equator are carried to the Polar Regions, and in this case the North Pole - warm surface water from the Bay of Mexico is carried northward to the Arctic, whereby it cools and the cold water sinks to a great depth to be carried southward to start the cycle all over again. If it were not for its mellowing effects on the climate the British Isles and north-western Europe would be a lot colder with places such as Scotland possibly never seeing the snow disappear during the summer months.

The Changeability of our Weather

To a large extent the weather we experience is determined by the particular kind of air which is being carried over our islands. The United Kingdom is situated at a meeting place for streams of air moving out of different regions and whichever happens to be the stronger at the time will determine our weather. During the winter months there are two very large High Pressure areas in the Northern Hemisphere, one over Siberia and the other over North America. When the Siberian High extends out over Scandinavia it will affect the flow of air across Britain, and together with the Arctic Ocean will produce very cold air over the United Kingdom.

Another area of persistent High Pressure is in the southern part of the North Atlantic Ocean and is known as the 'Azores High'. Due to its southerly oceanic position this area exports warm moist air - it is the origin of the type of air which commonly approaches the United Kingdom from the south-west. Between Greenland and Iceland there is an area of generally Low Pressure (the 'Icelandic Low') and it is from this region that the very cold damp air arrives in the United Kingdom. A similar conflict takes place during the summer months except the interior of the Continent is hot and dry, although moist air still approaches from the northerly and southerly parts of the Atlantic. The 'Azores High' is stronger and is found farther north in the summer due to the northward movement of the Sun - being overhead at the Tropic of Cancer in June; the 'Icelandic low' is still present at this time, but it is that much weaker.

Generally speaking the air approaching the United Kingdom from the southwest is warm and moist, the northwest it is cold and damp, north is cold, northeast is hot during the summer and very cold during the winter, and warm and dry from the southeast.

A significant factor affecting our weather is the Northern Jet Stream which plays a significant role when it comes to the weather across the

UK. It should be noted though that there are several Jet Streams around our planet, consisting of ribbons of very strong winds which move weather systems around the globe. They are found in the upper levels of Earth's atmosphere at the tropopause - the boundary between the troposphere and stratosphere, approximately 9-16 kilometres above the surface of the Earth, and can reach speeds of 200 mph. The position of the Jet Stream varies within the natural fluctuations of the environment. They are caused by the temperature difference between tropical air masses and polar air masses. What happens in one part of the world depends on what is happening elsewhere. The Northern Jet Stream can cause Atlantic depressions to deepen explosively as they are steered towards the UK, so they are very important in weather forecasting.

The above explanations are somewhat simplistic as all the forces which determine the weather over the United Kingdom are highly complex, it does however, help illustrate the reason for the variability of the weather the United Kingdom experiences.

Climate Change

Having discussed the weather we will now consider the climate recognising that the difference between weather and climate is simply a measure of time - weather is what conditions of the atmosphere are over a short period of time, and climate is how the atmosphere 'behaves' over relatively long periods of time. But before looking at climate change and global warming it will be helpful to understand what is meant by the Greenhouse Effect and what Greenhouse Gases are.

In understanding the Greenhouse Effect, it is useful to remind ourselves again of some basic physics, such as the fact that warm air will rise and that expanding air will cool. Thus, warm air near the surface of the Earth will rise, expanding and cooling due to the lesser pressure difference - on the other side of the coin compressing air warms it, as is easily observed when pumping up a bicycle tyre as mentioned earlier - the pump gets hot. From a logical point of view, the air MUST get hotter when compressed. Temperature is the key, being defined as a measure of the average kinetic energy, and that average results from collisions between molecules and the thermometer per unit time. So, as the gas is compressed, the distance between molecules decreases, which shortens the time between collisions, which mean the number of collisions with the thermometer increases per unit time, which means the temperature increases.

A phenomenon known as the Fohn wind is a good example of air being squeezed (compressed) in the atmosphere, and on 27th January 1958 at Abergwyngregyn on the North Wales coast, temperatures reached as high as 18.30 ^{0}C (65 ^{0}F), due to this effect, happening yet again on 10 January 1971 - at the time of writing it is the highest January temperature recorded in the United Kingdom.

The Garden Greenhouse

Although the term Greenhouse Effect is used to describe both the heating within a garden greenhouse as well as the atmosphere, it is important to fully understand the mechanisms at work in these two different situations.

The air inside a garden greenhouse is not subject to the same forces as the atmosphere in the sense that the air inside is trapped by the walls and roof of the greenhouse. Although convection takes place inside the greenhouse it is contained, and it is important to realise the air outside the greenhouse is subject to what can be called as 'uninhibited free convection', such that when warm air rises, cold air rushes in to replace it. Additionally, the volume of air inside a greenhouse is extremely small and simple when compared to the volume and complexity of the atmosphere.

The warming of the interior of a greenhouse takes place because solar radiation penetrates the glass and warms the soil, tables, pathways and plants inside. The longer (infrared) wavelengths emitted by the heated surfaces within the greenhouse cannot get out through the glass, and are trapped contributing to the heating effect within. This is what is known as the Greenhouse Effect: a transparent medium such as glass allows only incoming short waves to pass and blocks outgoing long waves. As a result, infrared radiant energy is trapped. Thus, the glass acts, in a sense, as a one-way valve, allowing visible light to enter, whilst preventing long waves (infrared) from leaving – being aware also that long wave-length infrared cannot penetrate the greenhouse.

The one-way valve effect is not the primary mechanism which heats the interior of the greenhouse. When solar radiation heats the ground inside the temperature of the ground rises. This increase in temperature is conducted to the air next to the ground, and then the air rises to the top of the greenhouse by means of convection. The 'trick' of course, is by trapping convection heat inside. As we have said earlier, the air outside is subject to uninhibited free convection, such that when warm air rises, cold air rushes in to replace it. Thus, the garden greenhouse situation, with its walls and roof, cannot apply to the atmosphere, which experiences free convection.

Obviously, to cool a greenhouse it is only necessary to simply open a vent, window or door to allow cooler external air to enter the greenhouse, thus convection being much more effective.

It is also instructive to remember how hot roof spaces can become due to lack of uninhibited free convection - many roofs do not have any windows (glass) and covered in slates or tiles. I can well remember the slated roofed house when I lived in Cardiff and how hot the internal roof space became during the summer months. This was due to the Sun's rays heating the external surfaces of the dark slates – heating the slates - and thus warming the air molecules touching the internal surfaces of the roof slates – and as a result the internal roof space heating up due to convection currents. Thus this convection heating process being exactly the same as that for the greenhouse, and indeed confirms the main mechanism that heats both glazed greenhouse and slated house roof.

Experiments have shown that when greenhouse glass is replaced by special glass that transmits long wave-length infrared light, the internal temperature is only slightly lowered – thus confirming that the absorption of infrared radiation is not the primary mechanism by which a greenhouse is heated, and it is important to realise that the inhibition of *free convection* is the main factor in warming a garden greenhouse.

To reiterate, the Greenhouse Effect (the nature of the properties of glass) is not the predominant factor which warms the interior. It is the lack of uninhibited free convection, and is easily proven by opening the windows or door - forced convection within the closed greenhouse, by means of a fan, would not prove very effective, simply because glass is a bad conductor of heat. Indeed, modern ovens have fans fitted and these are not provided to cool the Sunday roast down. Finally, it should be noted that convection warming action applies equally to central heating radiators. A person does not feel warm in a room as result of the direct radiation, but primarily because of convection currents. Thus, it can be argued that a more apt name for these radiators should be central heating convectors.

The Earth and the Greenhouse Effect

We are all aware the light and heat that we enjoy on our planet comes from our star, the Sun. This heat and light travels through the vacuum of space in the form of radiation - remembering that conduction, nor convection, can take place in a vacuum (space) - and we know that radiation is the transfer of energy at the speed of light by means of electromagnetic waves; this radiant energy lies mainly in the range of ultraviolet, visible light and short-wave infrared. Thus the Earth and its atmosphere are warmed by radiant energy from the Sun. The Earth, in turn, emits what is called terrestrial radiation, much of which is lost to outer space (69 per cent) – changes in the average temperature of the Earth are dictated by the difference in radiant energy entering and the radiant energy leaving. When the overall rates of energy gain and loss are the same then the atmosphere is said to be in *dynamic equilibrium*.

A good analogy would be that of a hydro-electric scheme: When the overall waters (streams and precipitation) entering the reservoir are equal to the waters being let out (to drive the generators) then the level of the waters behind the dam will remain constant, and thus the system is said to be in equilibrium. Approximately 49 percent of the incoming solar energy passes freely through the atmosphere and is absorbed at the Earth's surface, with an additional 20 per cent being absorbed by the atmosphere, this giving a total of 69 per cent being absorbed by the planet. The clouds reflect 22 per cent of the incoming energy, with an additional 9 per cent being reflected by the ground - if we add up all these figures they will account for all the incoming solar energy (100 per cent).

The long-wavelength infrared energy emitted from the surface of the Earth are not transmitted freely through the atmosphere – much of this terrestrial radiation is absorbed by water vapour and carbon dioxide in the atmosphere, and is radiated back to the Earth keeping the temperature higher than it would be otherwise. This trapping of the long-wavelength infrared is similar to the effect of the glass in a garden greenhouse and hence this why the overall process became known as the *Greenhouse Effect*. The radiation from the clouds and Greenhouse Gasses is emitted in all directions with 57 per cent escaping back into space. A certain amount of radiation (12 per cent) also escapes from the ground back into space. Therefore the total radiation escaping back into space is 69 per cent which is equal to the absorbed incoming solar radiation (69 per cent), and thus *dynamic equilibrium* is achieved.

But it should be clearly noted that the planet is considerably warmed by the Greenhouse Effect (trapping of long wave-length infrared), whilst garden greenhouses, (trapping convection heat inside), are not. I feel there is a certain irony about this. The amount of solar energy per second striking each square metre of the surface depends on time of day, season, atmospheric conditions and latitude. For example, because the Earth is a sphere, solar radiation striking the Polar Regions is spread over a larger area than that of Equatorial Regions, and we all know the effect of that. The amount of solar energy received each second over each square metre at right angles to the Sun's rays at the top of the atmosphere is 1350 Joules (1.35 kJ) - this amount of energy is called the Solar Constant - expressed in terms of Solar Power, it is 1.35 kilowatts per square metre (1.35 kW/m^2).

It is said that if the Earth had no Greenhouse Effect then the average surface temperature would be about -18 ^{0}C. However, the heat retention effect of the Earth's atmosphere, due to the Greenhouse Effect, means that *dynamic equilibrium* occurs at approximately 15 ^{0}C.

It could be argued that the Greenhouse Effect is not an appropriate term to use for the heating of *both* a garden greenhouse and the atmosphere, as free convection is nullified in the enclosed volume of a greenhouse, and the interior can only be cooled (vented) by opening a door or window – although in 2001 a NASA research team reported finding a huge climatic 'heat vent' in the atmosphere at the equator. This 'vent' appears to open naturally to release extra heat when the sea surface temperature rises. High clouds (Cirrus) over the western tropical Pacific Ocean seem to systematically decrease when the sea surface temperatures are higher, and a decrease in cirrus cloud cover will cool the planet by allowing more heat energy to leave the atmosphere - thus, the study suggests that our planet is much more active in managing its atmospheric temperatures than had been assumed, so perhaps, in view of this venting, the Greenhouse Effect is an appropriate term for both greenhouse and atmospheric heating after all.

Greenhouse Gas

Another common term bandied about these days is *Greenhouse Gas*, which is commonly accepted by many to apply only to carbon dioxide. But this is erroneous - what of the properties of water vapour or methane - water vapour is the most plentiful and important greenhouse gas that makes up most of the natural Greenhouse Effect, with CO_2 making up roughly half of the remainder.

The other greenhouse gasses are ozone (O_3), nitrous oxide (N_2O), methane (CH_4) and other trace gasses. In some circles CO_2 appears to have earned itself a bad reputation to the extent of being referred to as a pollutant. This is totally unjust and is a very good example of the misinformation perpetrated by ignorant and/or unscrupulous people. Without this gas, life would cease to exist on our planet. Indeed, all plant life depends on carbon dioxide, recognising that plant life is fundamentally at the bottom of the food chain, simply stated, herbivores eat plants, carnivores eat herbivores – without CO_2 we all die!

On the Continent, it is common practice to force (pump) carbon dioxide into greenhouses to encourage plant growth. In truth, carbon dioxide can only be considered a pollutant when it is found in great excess, such as on Venus, but then anything can become a pollutant when it is found in excess and upsets the natural balance. The Chambers Dictionary, with regard to pollution, offers the following definition, *'to make (any feature of the environment) offensive or harmful to human, animal, or plant life'*.

Global Warming advocates claim that the burning of fossil fuels releases more carbon dioxide into the atmosphere at a faster rate than plants can absorb it. But I will argue that a warmer Earth with more atmospheric carbon dioxide (higher level of temperature for *dynamic equilibrium*) would make the planet a more comfortable and enjoyable place to live – trees and plants would certainly benefit from these conditions, and in turn, we Humans as well - they pump CO_2 into greenhouses in Holland not to poison the plants, but on the contrary, to enhance the plants and increase crop yields.

Anthropogenic Global Warming (AGW)

Commonly known as man-made global warming - this has yet to be proven as the main driving force for any atmospheric warming, although man-made carbon dioxide is obviously an addition to the atmosphere.

This addition to the atmosphere might be the cause for an increase in any global warming, but it is just not proven, regardless of what parts of the media and some climatologists would claim. There are many other significant forces and mechanisms that are responsible for climate change, that's if indeed, the Earth is headed for significant global warming and not global cooling, and perhaps a little ice age – after the freezing winter of 1962-3 there was a lot of talk about another ice age – now it appears we are going to cook?

Political obfuscation lends itself to all the confusion coupled with a profusion of scare stories such as the claim that the sea ice at the Arctic is decreasing – the Northwest passage will be open all through the year - triggering silly stories in the media of polar bears drowning due to lack of ice – whilst at the same time the sea ice in the Southern Hemisphere (Antarctic) is increasing, as determined by research done by the Polar Research Group, Department of Atmospheric Sciences, University of Illinois, USA.

In reality the climate is bound to change simply because we live on a dynamic planet, subject to numerous changeable natural forces as history and the geological record will testify, and it will be instructive to look at some of the natural forces and mechanisms which bring about climate change - hopefully the reader will then  have a better understanding of just how complex the subject is, and thus be able to see through some of the rubbish that a number of people spout in pursuance of their own immoral and myopic agenda – let us hope that common sense and true knowledge will prevail.

As mentioned earlier in this chapter, the atmosphere is a mixture of gasses that surround our planet - since these gasses are fundamentally made up of atoms, which have mass, it follows that the atmosphere has mass and is therefore prevented from escaping into space by the gravitational force of the Earth.

It should be noted that planetary bodies with weak gravitational fields lose their atmospheres fairly quickly.

As an example, Mars has a very *tenuous* atmosphere, with the pressure at the surface less than 1 per cent of sea-level pressure on Earth. The Martian atmosphere is composed of more than 95 per cent carbon dioxide, with the balance consisting mainly of molecular nitrogen and argon – with traces of water vapour, oxygen and carbon monoxide. Martian weather is different to that on the Earth, with dust storms capable of covering all or most of the planet.

Our Moon with its low mass and hence weak gravity has no atmosphere and hence no weather. There is only a narrow zone that supports life on Earth and it is called the Biosphere – it is limited to the waters of our planet, a fraction of the crust, and the lower regions of the atmosphere.

The Biosphere, by its very nature, is a dynamic system subject to various influences which change atmospheric, sea and land temperatures over time. The climate of the Earth changes over the short, medium and long term. Short term climate changes are obviously the seasons, such as in the northern hemisphere, experiencing spring, summer, autumn and winter; these changes being due to the inclination of the Earth's axis and orbit around the Sun. Indeed, it may be argued that a very short climate change (not to be confused with the weather) is the difference in temperature between day and night in, for example, a desert region – extremely hot during the day and very cold at night.

Looking at the other end of the scale, a long term example of climate change would be the major ice-ages, known as glacials. The periods between glacials and warmer times, known as interglacials, are hundreds of thousands of years, with the last major ice age reaching its maximum about 20,000 years ago and ending about 12,000 years ago. It should be noted that the Earth does experience less severe cooling during the interglacials, and these are called Little Ice Ages. One such Little Ice Age occurred as relatively recent as the 17th Century, as paintings in the Museum of London will testify. An example being the River Thames in central London in 1677, showing stranded ice floes by Abraham Hondius (1630-1695); such cooling events like the Little Ice Age have occurred nine times since the last major ice age (glacial) ended.

The forces which contribute to weather and climate change, are many and extremely complex, and a number of known mechanisms, in no particular order of merit, are explained as follows:

El Niño Effect

The name, translated from the Spanish, means 'the little boy' as it occurs usually around the Christmas season, and refers to the Christ child. The El Niño effect occurs roughly every seven to fourteen years, and can have a large effect on global weather. The effect is temporary, but can cause severe atmospheric and oceanic change in the climate of the Pacific Ocean, in an area around the equator. The El Niño effect can lead to a complete reversal of the trade winds, resulting in torrential rain fall, culminating in flooding and mudslides to the usually dry coastal regions of Peru and Ecuador in South America. It can also result in the collapse of the monsoons in Asia, bringing severe drought to places such as Indonesia and Northern Australia; severe weather disturbances can also occur in other parts of the globe, such as droughts in areas of Africa and central North America. It is interesting to note that the effects of El Niño in Peru can be found in written records that go back to the sixteenth century, and researchers have found geological evidence of El Niño in Peruvian coastal communities from at least 13,000 years ago. Indeed, it is claimed the Inca were also aware of the effect. The tropical Pacific is now continuously monitored by satellites and about 70 buoys in what is known as the Tropical Atmosphere-Ocean Array (TAO). For those readers who wish to delve further into this topic, they should know that the El Niño phenomenon is part of a larger system known as the El Niño Southern Oscillation (ENSO), which was discovered in 1923.

La Niña Effect

Translated from the Spanish this means, 'the little girl' and is an abnormal cooling in the eastern Pacific producing conditions more or less the opposite of those created by the El Niño effect, and are most noticeable from December to March.

Sunspot Cycle

Sunspots are relatively dark areas on the bright white surface (known as the photosphere) of the Sun - they are usually between 1,000 and 40,000 kilometres (1,609 miles and 64,360 miles) in diameter and are caused by strong magnetic fields that mark areas of heightened magnetic activity on the photosphere - Sunspots look dark because they are cooler, $3,500\ ^0K$ (degrees Kelvin), than the surrounding bright surface of the Sun at $5,800\ ^0K$. The very small spots of just a few hundred kilometres across, may last for only a few hours, whilst the very large ones can last for months - Sunspots are not constant in number, increasing and decreasing over a period of about eleven years.

Historical records indicate long periods when there appeared to be essentially no Sunspots at all. This phenomenon was noted by two astronomers, firstly by the German astronomer Gustav Spörer (1822-1895) and the British astronomer E.W. Maunder (1851-1928). In their honour we now have the Spörer minimum (1450-1540) and the Maunder minimum (1645-1715) which was confirmed by radiocarbon dating of tree rings, as being times of low solar activity. It was noted that these minima coincided with prolonged periods of cold weather, suggesting the Sun's significant influence on the Earth's climate - we have seen earlier that a little ice age occurred during the 17th Century – just when Sunspots were at a minimum.

At the time of writing this chapter the current Solar Cycle 24 has not been as active as Solar Cycle 23 although the observed monthly Sunspot number increased during 2013/2014 with over a 100 spotted during February 2014. Solar Cycle 24 began on 8 January 2008 and is the 24th solar cycle since 1755, when recording of solar Sunspot activity began - it should be noted that there was minimal activity through early 2009. Predictions are that Solar Cycle 24 could be the least active cycle in the past one hundred years, and judging by the historical record this failing would suggest we are heading for a cooling period of the climate, rather than a warming. In relation to solar activity and Sunspots see also Coronal Mass Ejections (CMEs) later in this chapter.

Cloud Formation and Cosmic Rays

Highly energetic charged particles known as cosmic rays constantly bombard the Earth from space, and as such they are continually raining down on us. Low-energy cosmic rays come from the Sun whilst higher energy particles come from sources within our Galaxy, such as supernovae. For readers not familiar with astronomical terms, a supernova is the explosive death of a star that is so violent that for a brief period the star shines as brightly as a whole galaxy of more than 100 billion stars similar to our Sun, and is a relatively rare event.

The highest-energy particles probably come from beyond our Milky Way Galaxy. Before cosmic rays can reach the surface of the Earth they have to overcome three defensive shields in the form of the Sun's magnetism, the Earth's magnetism (any weakening of these magnetic fields will allow more cosmic rays through) and the atmosphere. The primary cosmic rays collide with molecules in the atmosphere to produce large numbers of other particles namely, secondary cosmic rays which in turn decay to produce an extensive air shower covering several square kilometres of the Earth. How many people appreciate that a cosmic ray particle travels through their body approximately twice every second before disappearing into the ground under their feet? To discover more about the effect of cosmic rays and how they affect cloud formation I would recommend the book titled, 'The Chilling Stars (A Cosmic View of Climate Change)' by Henrik Svensmark and Nigel Calder.

Milankovitch Cycles

The Serbian astrophysicist Milutin Milankovitch (1879-1958) developed a theory relating Earth motions such as orbital eccentricity, changes in obliquity and precession, which together have become known as Milankovitch cycles. The theory explains long-term climate change over periods of thousands of years; a 1976 study which examined deep-sea sediment cores found that when the Earth was going through different stages of orbital variation around the Sun, ice ages had occurred.

Orbital Eccentricity

The shape of the Earth's orbit around the Sun is determined by its eccentricity, where the closer the eccentricity is to zero (0) the closer the shape is to a circle: the eccentricity of the Earth is 0.0167 and the periodicity of changing from less elliptical to more elliptical is about 93,000 years. It is interesting to note that the shape of the planetary orbits of the Solar System tend more towards the circular than the elliptical, with the exception of Mercury at 0.2056 and Pluto at 0.2484. Due to the elliptical shape of its orbit, the Earth is further from the Sun in July than in January, receiving 7 per cent more light and heat in January than July - as these dates roughly correspond with the solstices we find that, on the whole, the northern winter, occurring when the Earth is nearest to the Sun, tends to be less rigorous than the southern, which occurs when the Earth is receiving minimum light and heat. The northern summer is not as warm as the southern when the radiation received is at its maximum. The nearest point the Earth is to the Sun is called perihelion, whilst the most distant point from the Sun is known as aphelion.

Obliquity of the Ecliptic

Obliquity is the angle between the equatorial and orbital plane of a body, which can be defined as the angular distance between the rotational and orbital planes. Therefore in relation to the Earth, obliquity of the ecliptic is the angle between the planes of the equator and the ecliptic, where the ecliptic is the apparent path of the Sun around the sky. This change in the inclination of the Earth's axis (axial tilt), which varies (from 21.8 to 24.4 degrees) occurs over a period of about 41,000 years; currently the inclination is 23.5 degrees. Thus the angle at which the rays of the Sun strike the Earth's surface varies with latitude and season. As an example the Sun is highest in the sky at noon during the summer solstice (about 22nd June) in the northern hemisphere, and lowest in the sky at the winter solstice (about 22nd December). The opposite is true for the southern hemisphere as at the vernal and autumnal equinoxes, day and night are equal there, whilst at the equator the Sun is more or less directly overhead at noon year round. Also due to this inclination the Sun at the poles, although low in the sky, never sets for six months - nor rises above the horizon for six months.

Precession

Precession is the change in the direction of the Earth's axis of rotation. The axis of rotation behaves like a spinning top and as such traces a circle on the celestial sphere over a period of time - a wobble of the Earth's axis taking almost 26,000 years for each cycle. Currently the axis points towards Polaris (the North Star), but in 13,000 years' time it will point towards Vega. This is the brightest star in the constellation Lyra, the fifth brightest star in the night sky and the second brightest star in the northern celestial hemisphere. It is a relatively close star at only 25 light-years from Earth; Vega was the northern pole star around 12,000 BCE.

Dansgaard-Oeschger Cycle

This is a moderate, irregular 1,500 year solar-driven cycle that imposes its effect on most of the Earth's fairly-constant climate fluctuations. Three scientists led to the discovery of the 1,500 year cycle, namely, Willi Dansgaard of Denmark, Hand Oeschger of Switzerland and Claude Lorius of France. They were jointly awarded the Tyler Prize (environmental Nobel) in 1996 – although their award citations never mentioned a word about the climate cycle they discovered, or anything about its predictive power to forecast moderate climate changes.

No doubt the disciples of anthropogenic warming will be in denial to the Dansgaard-Oeschger Cycle, and the many other cycles that have an effect on our climate, so it would be enlightening for them to familiarise themselves with the Roman Warming Period (200 B.C. to 600 A.D.) - a visit to the Roman city and vineyards at Wroxeter, Shropshire would offer further educational insights – these disciples should also research the Medieval Warming (900 A.D. to 1300 A.D.) and the little Ice Ages (1350 A.D. to 1850 A.D.) when the Thames froze over and offered the opportunity for winter fairs on the ice - a little 'Nordic' reading will also 'broaden' their horizons to learn of the world of Eric the Red and the Greenland colonisations – have they, I wonder, ever considered why Newfoundland was once called Vinland - climate change has always, and will always affect the planet, regardless of the presence of Homo sapien and his works.

The following indicates, from the historical record to the present day, the climate change in relation to warming and cooling:

PERIOD	DATE
Roman Warming.............................	200 B.C. – 600 A.D.
Dark Ages Cooling...........................	440 A.D. – 900 A.D.
Medieval Warming..........................	900 A.D. – 1300 A.D.
Little Ice Age, Phase 1 & 2.................	1300A.D. – 1850 A.D.
The Modern Warming (Global Warming).	1850 A.D. – Present Time.

The periods of climate change indicated above are indisputable facts, unlike the modern day computer models, which fundamentally can be anyone's guess – we all know how accurate meteorologists are at forecasting the weather – The Great Storm of 1987 (denied on UK television prior to the storm) was a hurricane that occurred on the night of 15–16 October, 1987 - with ferocious winds causing casualties in England, Channel Islands, and France as a severe depression in the Bay of Biscay moved northeast - we are all aware of barbecue summers that never materialise, and yet climatologists warn of global warming, when meteorologists cannot accurately forecast the weather for a six months ahead!

The climate model enthusiasts should keep in mind the very real truth relating to computers generally: garbage in, garbage out (GIGO). A book that I would certainly recommend regarding the 1,500 year climate cycle is called 'Unstoppable Global Warming' by S. Fred Singer and Dennis T. Avery. Indeed, the book became a New York Times best-seller, and stayed on that distinguished list for several months.

Volcanoes

Volcanism has an important role to play in climate change - 250 million years ago volcanic eruptions decimated over 90 per cent of life on Earth in what has become known as the great extinction. Eruptions inject small, sulphate particles into the atmosphere – this has the effect, whilst they last, of reflecting solar energy back into space and cooling the Earth. The Mount St. Helens volcano, Washington State, United States, during 1980 made headlines when it erupted spewing out over 1 cubic kilometres of material, but this eruption was relatively small in historical terms – the eruption of Vesuvius, Italy, in 79 BC was about five times as big, ejecting out some 5 cubic kilometres of material. Krakatoa, Indonesia, when it erupted in 1883 A.D. produced twice as much again and had a Volcanic Explosivity Index (VEI) of 6. The VEI can be regarded as a volcanic Richter scale and it is also logarithmic, which means that each point on the scale denotes an eruption 10 times larger than the one below. For example Mount St. Helens had a VEI of 5, whilst Krakatoa came in at 6, and its ash, in the upper atmosphere, had truly remarkable effects on sunsets worldwide.

But for a real 'corker' and for something of a very different order of magnitude we need only go back to 1815 A.D. when Tambora, Indonesia ejected 100 cubic kilometres of material into the atmosphere bringing global cooling and severe crop failure for at least three years; this had a VEI of 7. Unfortunately and disconcertingly, there have been far greater eruptions in the past - these gargantuan eruptions being due, to what are known as Supervolcanoes, such as the Yellowstone Caldera, in Yellowstone National Park, Wyoming, North America - and can have a devastating and long lasting effect on global temperatures. It should be noted that the three largest eruptions in North America during the past few million years all occurred in Yellowstone! The last eruption occurred about 640,000 years ago with an estimated VEI of 8, and it is claimed we are now due for another one – although the chances of a super-eruption is given as about 1 in 700,000 in any single year. The last super-eruption occurred 73,500 years ago when Toba, Indonesia erupted with an estimated VEI of 8. This eruption triggered a cold period that lasted for five or six years; the effects being confirmed in ice-cores records from Greenland. Hard freezing occurred in temperate regions such as Europe and North America. It has been suggested there was a human population crash with just a few thousand people surviving the period of bitter weather.

Other Influences

Over the extremely long term we have the passage of the Solar System (passing through dust clouds for example) around the Galactic centre and plate tectonics whereby the land masses move over the surface of the planet from warm to cold areas, and vice versa, altering ocean currents et cetera. The historical and geological record will clearly show how the climate has changed in the UK and across the globe.

Coronal Mass Ejection (CME)

Having discussed the various influences that dictate our weather and climate on planet Earth it would be remiss not to mention space weather and how the Sun has had, and can have, a detrimental effect on not just modern communication links but electrical distribution systems as well. This damaging scenario comes in the form of a Coronal Mass Ejection (CME) from the Sun, which is created from a solar storm whereby a massive burst of solar wind and magnetic fields rise above the solar corona and are released into space. Most ejections originate from active regions on the Sun's surface, such as groupings of sunspots associated with frequent flares. Solar maximum (solar maxima) is a period of greatest solar activity in the 11 year solar sunspot cycle.

During solar maximum, large numbers of sunspots appear and the Sun's irradiance output grows by about 0.1 per cent. The increased energy output of solar maxima can impact global climate and recent studies have shown some correlation with regional weather patterns. At solar maximum, the Sun's magnetic field lines are the most distorted due to the magnetic field on the solar equator rotating at a slightly faster pace than at the solar poles. The solar Sunspot cycle takes an average of about 11 years to go from one solar maximum to the next, with an observed variation in duration of 9 to 14 years for any given solar cycle.

Large solar flares often occur during a solar maximum, indeed the solar storm of September 1859, (during Solar Cycle 10), the largest recorded geomagnetic perturbation, known as the Carrington Storm after the English amateur astronomer Richard Carrington (1826-1875), struck the Earth with such intensity that the northern lights could be seen as far south as Rome. CMEs can cause particularly 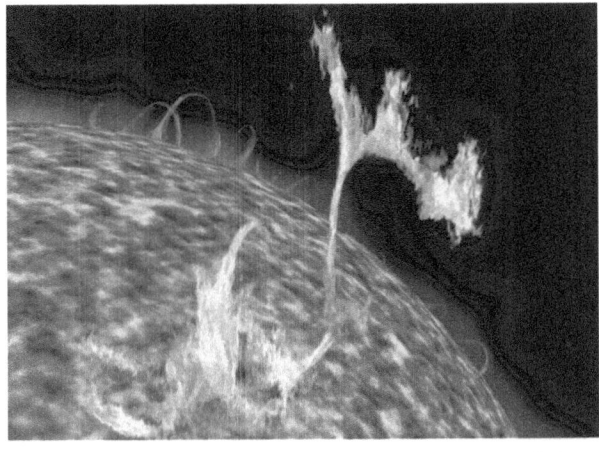 strong aurorae and can have a dramatic effect on the Northern Lights (Aurora Borealis) and the Southern Lights (Aurora Australis).

During the storm of 1859 telegraph operators suffered electrical shocks, indeed the storm took down parts of the recently-created US telegraph network, starting fires and as mentioned, shocking some telegraph operators. Coronal mass ejections, along with solar flares of other origin, can disrupt radio transmissions and cause damage to spacecraft electronics, and increase drag on satellites so they consume more fuel to maintain their proper orbits. But the most serious potential for damage are electrical transmission lines, resulting in potentially massive and long-lasting power outages. Solar storms, and particularly CMEs, can pose a health threat to astronauts in space or airline passengers passing over the poles, where protection from Earth's magnetic field is at its weakest. The most worrying aspect of CMEs is that the 1859 solar storm observed by the astronomer Richard Carrington begs the question of when will the next big event take place?

Somewhat frighteningly NASA's Solar Terrestrial Relations Observatory-Ahead (STEREO A) spacecraft detected a coronal mass ejection on the Sun on July 22, 2012. Had this eruption occurred nine days earlier, it would have hit the Earth, with the potential to wreak havoc with the electrical grid, disabling satellites and GPS, and disrupting our increasingly electronic lives - the solar bursts would have enveloped Earth in magnetic fireworks matching the largest magnetic storm ever reported on Earth – namely the aforementioned Carrington event of 1859.

The STEREO (Solar Terrestrial Relations Observatory) team based at UC Berkeley's Space Sciences Laboratory said that if the storm had hit the Earth, it probably would have been like the big one in 1859, but the effect today, with our modern technologies, would have been staggering – a recent study estimated that the cost of a solar storm like the Carrington Event could reach $2.6 trillion worldwide.

It has been suggested that there is about a 10 per cent chance of another intense storm in any particular year - no one really knows how bad a storm such as the 1859 event could be – remember before the 19th century there were no electrical networks across the face of the planet – today we are totally dependent on our power and communication networks. Everything from banking to buying groceries depends on electrical energy in its many form of usage – imagine the sudden shock to society if the electrical power *suddenly failed* coupled with telephone landlines and the Internet going down – we are so reliant on electronic transactions for everything - everyday items such as fuel for the car would become unobtainable, railways dependent on electrical power would fail, people would be unable to get cash from the bank, satellite communication and TV would fail.

But we just don't know for sure *how serious* the situation would be should the Earth experience a storm equal to, or indeed, more powerful than the Carrington event – it is sobering to remember that during March 1989 (during Solar Cycle 22) a Solar storm approximately a third less powerful than the Carrington event caused the collapse of the Hydro-Quebec electricity transmission system. Due to the very long transmission lines of the Hydro-Quebec power grid, coupled with the fact that most of Quebec sits on a large rock shield (high ground resistivity), these factors inhibited current flowing to ground (earth) and as a result the electrical current finding a less resistance path along the 735 kV power line, which as a consequence tripped out circuit breakers. This resulted in more than six million customers losing their power for up to nine hours. Much further afield a number of polar orbital satellites lost control for several hours and the GOES weather satellite had communication problems resulting in the loss of weather images.

The reader should recognise that CMEs consists of a billion tons of matter traveling at a million miles an hour through space, and although these storms are huge and powerful, they are very tenuous and widely dispersed with just a few particles per cubic centimetre such that much of the power they have to affect us comes from their magnetic fields – geomagnetic storms - but at least with current space technology and the Sun being 150 million kilometres (93 million miles) from the Earth we would have some warning – it takes light only 8 minutes to make the journey to Earth, it usually takes a CME two to four days, but extremely fast ones have been known to reach here in just over a day.

So what if the Earth were to experience an extreme space weather storm (a Solar Superstorm), which is a low-probability, high-consequence event that poses severe threats to critical infrastructures of our modern technological society. It was concluded that the huge outburst on the Sun on July 22, 2012, propelled a magnetic cloud through the solar wind at a peak speed of more than 2,000 kilometres per second, four times the typical speed of a magnetic storm. It tore through our planets orbit but, luckily for the Earth and the other planets which were on the other side of the Sun at the time, salvation ruled the day. Any planet with a magnetic field in the line of sight would have suffered severe magnetic storms as the magnetic field of the outburst tangled with the planets' own magnetic fields.

One reason this particular event was deemed so potentially dangerous, was apart from its high speed, is that it produced a very long-duration, southward-oriented magnetic field. This orientation drives the largest magnetic storms when they hit Earth because the southward field merges violently with Earth's northward field in a process called reconnection. Storms that normally might dump their energy only at the poles instead dump it into the radiation belts, ionosphere and upper atmosphere and create auroras down to the tropics.

Hopefully when the next big event arrives, as it surely will, we will be able to cope - if not, it could be Armageddon – during October, 2014, a large Sunspot AR12192 with a diameter of 129,000 kilometres, the largest for a quarter of a century, unleashed six enormous X-ray flares that ionised the Earth's upper atmosphere – but when the big event does occur the Earth could be safely on the other side of the Sun.

Writing about CMEs and their potential threat to life on Earth reminded me of my Sunday School lessons many, many years ago – and in particular the story of Noah and the flood when life on Earth, apart from that on board the Arc, was destroyed when the Earth was deluged with rain for forty days and forty nights, causing a great flood, (Genesis 7) - but also of the future destruction of the Earth by fire where it is written in the Bible and I quote, *"They deliberately ignore this fact, that by the word of God heavens existed long ago, and an earth formed out of water and by means of water, through which the world that then existed was deluged with water and perished. But by the same word the heavens and earth that now exist have been stored up for fire, being kept until the day of judgment and destruction of ungodly men"* end of quote, 2 Peter 3: 5-7.

Summary

Weather and climate change are like handling slippery fish in attempting to grasp the reality and truth of it all – one needs to take a deep breath and apply a little common sense in trying to understand the forces of nature and the impact of Human society on our atmosphere. To claim that our modern society cannot have any effect upon the large environment we call Earth is being perhaps a little naïve. The use of an analogy can be useful insomuch that if we imagine a bath tub fall of fresh water into which we introduce just a grain of salt, we would not expect the bath water to taste salty. But if we were to continue to introduce additional grains of salt, say at the rate of one grain a minute, then at some stage we would expect to detect a change in the bath waters salinity - the point to note is that there will indeed come a time when a change is detected – this time being directly related to our sense of taste, the volume of water, size and salinity of the grain of salt, and the rate of input.

The oceans cover most of the planet, being vast and deep containing almost innumerable life forms. Now who would have thought the vast numbers of cod could be devastated by over-fishing in the waters of the Grand banks, off the coasts of Labrador and Newfoundland. For sure, the Portuguese and Basque fishermen who worked the Grand Banks as early as the 15th century would have deemed it impossible – the world's richest fishing grounds - but, due to human failing (greed, lack of foresight and conservation) it happened! Closer to home in the North Sea it was thought that cod stocks were limitless, but over-fishing of the North Sea Cod stock resulted in a near collapse in 2001, although recent reports suggest there may be a slight recovery – time will tell? Bluefin tuna, which shoaled the Atlantic in vast numbers, were thought to have declined by 50 per cent during the 1960s, but now the figure is nearer 80 per cent. There are many other fish stocks that have been over-fished world wide - the oceans may be vast but they are not in any way infinite.

Similar events have happened on land and the decimation of the North American Bison (buffalo) is a good example – millions of wild buffalo once roamed the North American Continent from Mexico to Canada long before people settled there. The coming of the Native Americans who relied on the buffalo for practically everything from food and clothing to shelter did not have any meaningful impact on buffalo numbers. But later when the European settlers arrived in America things began to change – the Native Americans learned to use horses thus expanding their hunting range and enabling more buffalo to be killed. But it was only when white trappers and traders introduced guns the killing rate began to spiral resulting in the slaughter of millions of buffalo – unbelievably even train passengers were shooting buffalo for sport during the 19th Century – I can still remember reading my comics as a small boy and being enthralled to the adventurous tales of Buffalo Bill (William Cody) never realising this man, who was hired to kill buffalo, slaughtered more than 4,000 buffalo in two years – such is the innocence of children.

Currently I believe there are there are between 150,000 and 200,000 bison throughout North America, although the vast majority of them are raised on ranches for commercial purposes (mostly for meat, hides and skulls).

It is shameful to recognise that there is not a single area, from pole to pole, whereby human intrusion, exploitation or contamination can be denied. It is truly outrageous that nearly 400 marine species are at risk due to the tons of shopping bags, fishing nets and other waste dumped in the seas every year – plastic bags should be banned and replaced with recyclable (papers) bags as of yesterday. Due to apathy, ignorance and selfishness the Loggerhead Sea Turtle, North Atlantic Right Whale, Hawaiian Monk Seal, Shearwater, African Penguin and unbelievably the beautiful Albatross could be wiped out as a direct consequence of our pollution – every time dear reader you buy a plastic bag and discard it, think where it might end up – we can all help by refusing plastic bags, asking for the paper equivalent and lobbying our MPs. It is brilliant that the Daily Mail has been campaigning since 2008 for tougher action to reduce the problem, and in a major victory in 2014 it was announced that supermarkets and large stores in England will charge 5 pence per plastic bags (as they do now in Wales) from October 2015. But the final victory will only be when plastic bags are banned in law.

So it should come as no great surprise that we do have the ability to 'affect' the atmosphere, not just locally but even on a global scale. Most folk are aware that there is a temperature difference between town and country - it is warmer in the cities than the countryside in both summer and winter – the urban heat island effect. All the concrete, roads and buildings of an urban environment absorb solar radiation and give out heat. It is useful to remember that electrical night storage heaters convert electrical energy to heat energy by taking advantage of cheap rate night time electricity - this energy being used to warm up bricks inside the storage heater and then during the day give off this stored heat in a controlled rate to warm the room - thus a city can be seen, in this sense, to be a very large storage heater, but of course there is no containment and control knob to regulate the amount of heat radiated.

The city absorbs and radiates heat during the day as city dwellers know only too well, but cities and towns also radiate heat at night. This warming of urban environments by solar radiation is further enhanced by the artificial heat being generated in all the buildings from lighting (much less with LED lighting), but much more significantly, during the winter, from central and other forms of heating. Additionally, we must also take into account the population density as we mammals generate heat, and of course not forgetting the heating from the internal combustion engine - all these sources of heat contribute to what is known as Urban Heat Islands. Now it can be argued that currently a small number of these urban heat islands are nothing to get too hot under the collar about (pun intended) and will have little, if no effect on global atmospheric temperatures. But what has to be considered is how the spread and enlargement of many urban heat islands will it take before they do indeed begin to have an effect, on not only inter-urban island temperatures – but on truly global temperatures as well?

Additional warming will be beneficial up to a point, but just in the same context as light pollution it can all get out of hand as any astronomer will tell you – can you see the heavens and the multitude of stars from where you live, and have you ever seen the beautiful Milky Way from your urban heat/light island – I doubt it very much - you only have to look at a night-time satellite image of our planet to see how light pollution has spread dramatically across the face of the planet.

At the time of writing the Hadley Centre for Climate Prediction and Research (part of Britain's Met Office) has conceded that recorded temperature figures for the first seven years of the 21st century reveal there has been a standstill – they state on their website and I quote, "That over the last ten years, global temperatures have warmed more slowly than the long-term trend. But this does not mean that global warming has slowed down or even stopped. It is entirely consistent with their understanding of natural fluctuations of the climate within a trend of continued long-term warming. These natural fluctuations include the El Niño Southern Oscillations (ENSO) in the Pacific Ocean." End of quote.

As mentioned earlier and worth reiterating, this is an amazing amount of faith to put in advanced, state-of-the-art, computer modelling. Surely it is recognised that knowledgeable and sensible people are fully aware of the old maxim relating to computers: garbage in, garbage out (GIGO) - will they be telling us next that their marvellous computer modelling will reliably predict the weather for next summer - if this is so, then bring it on I say, as it will be nice to have a reliable forecast for a change.

Regarding the reliability of predictions during April 2009 the Met Office stated the UK was odds on for a barbecue summer – this turned out to be an inaccurate forecast as July and August turned into washouts – then during the same year forecasters said we could expect a milder than average winter, with only a one in seven chance of a cold Christmas season – in the event it turned out to be the coldest and snowiest winter for more than three decades. During March 2012 the Met Office predicted 'drier' than average conditions for April to June, and the quarter turned out to be the wettest since 1910 with widespread flooding. The Met Office has now stopped issuing long-term forecasts to the public and instead it continues to give 'probability' guidance for coming months to Government departments such as the department for Environment, Food and Rural Affairs (Defra) which needs to plan...but I guess the owners of wind farms and the National Grid are not going to hold their breath as to where and when the wind will be forecasted to blow next.

Therefore dear reader you will not be surprised when I say that it is not good science to claim global warming is *definitely* caused by man-made emissions at this moment in time, and that there is a massive body of evidence supporting this claim. There may be a substantial body of support for this assertion in parts of the media, and political arena, but there is certainly not a massive body of scientific evidence to support the 'anthropogenic' warming hypothesis as yet.

We have seen there are many other more significant forces at work that can explain climate change. The reader should recognize that historically an increase in atmospheric carbon dioxide has followed warming, and not the other way around. It is a fact that as the oceans heat up they release carbon dioxide. The followers of man-made global warming should recognise all the various forces and mechanisms that drive the climate have been in play long before the Industrial Revolution, or indeed the appearance of Homo sapien - they need also to be aware of the relatively new field of science called Cosmosclimatology.

Sensible climate scientists admit the Earth's climate is determined by hugely complex systems, and reliable prediction is difficult, if not impossible. The reader should be absolutely clear that I am not claiming Human's do not have the capacity to have an impact on our planets global temperature – I am simply saying that scientists just do not know for sure if global temperatures, on the long scale, are on the increase and due to human interference. But if temperatures are, then the true recognition of the force or forces which are responsible, have yet to be determined.

Hopefully this chapter has offered insights to the climate and that the current data tends to suggest that we could possibly be heading for a much cooler period, which in its turn, could negate any possible additional heating as a result of our species. Hopefully the reader will now not be surprised when I state that wind generation in the UK will have the same effect on CO_2 emissions and global temperatures as attempting to increase the salinity of the oceans with a spoonful of salt, as the next chapter will further reveal.

Concluding this chapter on weather and climate it is useful to record the following words:

The Independent newspaper on March 20th, 2000, had an article headed, "Snowfalls are now just a thing of the past." Where it quoted Dr David Viner, head of the climate unit at the University of East Anglia, saying that in future: "Children just aren't going to know what snow is."

I will leave the reader to draw their own conclusions to this offering bearing in mind there has actually been no global warming for the last 15 years and during 2012/3 the UK suffered its coldest winter with Siberian winds, blizzards and deep snow for 30 years with many hill farmers counting the cost of dead sheep and lambs!

Remember Al Gore stated that the Arctic would likely be ice-free in summer by 2014 – when in fact Arctic ice has been recovering lately and is likely to do so for several more decades - Al Gore needs to look in the mirror whilst remembering his Bible and Psalm 146:3 *Do not trust in princes, In mortal man, in whom there is no salvation. His spirit departs, he returns to the earth; in that very day his thoughts perish....*

Non semper ea sunt quae videntur

(Things are not always what they appear to be)

CHAPTER THREE

Wind Technology in the UK

Large scale wind technology offers no reduction in carbon dioxide emissions, causes environmental damage, is a threat to wildlife, livestock and produces invasive noise and flicker, requires government subsidies and backup by conventional power stations – it is not even effective in producing meaningful electrical power – just an absolute nightmare.

But before considering in detail the case against large scale wind generation of electricity, it is useful to compare the criteria required for the placement of a solar panel array as a means of producing electrical energy. Obviously it makes good sense to employ solar technology in the correct geographical location for maximum effectiveness. The setting up of large solar arrays in places such as Alaska, Greenland, Iceland, Scandinavia, Finland, or the Arctic and Antarctica would be foolish for the obvious reason of minimal solar radiation. Indeed, the Arctic and Antarctica have six months of weak sunlight (daylight) followed by six months of darkness as every school person knows. Common sense demands that to get the very best from large solar arrays they should be placed in areas of intense solar radiation such as around the equator, the sunny deserts of North and South America, Australia, Africa and Asia.

Nevertheless, if sensibly deployed solar panels can contribute quite effectively to power savings for office and domestic properties in parts of the UK. It is not 'rocket science' to conclude the deployment of solar arrays need careful planning and locating to realise their maximum potential – a north facing roof is not a good choice – and the further south of the country, the better.

The chapter headed, 'How to Reduce your Domestic Energy bill' explains just how effective solar panels can be for domestic users in the UK coupled with a remarkable reduction in mains power usage, and of course the power bill – so if nothing else, and to reiterate, the information offered will more than repay the cost of this book many times over especially in these days of our ever increasing energy bills.

Returning to the subject of wind generated electricity, it should be recognised that the wind has been employed for over two thousand years for driving such things as stone wheels to grind corn, or to pump water - indeed windmills were used in the East in ancient times; in Europe they were first used in Germany and the Netherlands in the 12th century. So perhaps it should come as no surprise, in this respect, that these same countries have a proliferation of electrical generators driven by wind power. According to the European Wind Energy Association (EWEA), at the end of 2008 Germany had the largest wind capacity at 23,903 MW, with Denmark at just 3180 MW, and the Netherlands 2225 MW. It is interesting to note that Spain and Italy exceed the wind generation capacity of Denmark and the Netherlands at 16,740 MW and 3736 MW respectively – have the Danes and the Dutch learnt something about wind generation that restricts their capacity - food for thought?

The EWEA also stated that at February, 2014 there were 117.3 GW of installed wind energy capacity in the EU: 110.7 GW onshore and 6.6 GW offshore - 11,159 MW of wind power capacity (worth between €13 bn and €18 bn) was installed in the EU-28 during 2013, a decrease of 8 per cent compared to 2012 installations.

At the beginning of 2014 the wind generator number for the UK stood at 5,276 offering a capacity of over 10 GW - according to the Department of Energy and Climate Change (DECC) the onshore capacity for wind generation was 7513 MW and offshore at 3696 MW offering a total of 11,209 MW – the actual generation was, 16992 GWh for onshore and 11,441 GWh for offshore, giving a total of 28,433 GWh.

The reader should fully recognise that although certain countries have a proliferation of wind driven generators, it does not signify or qualify a 'universal qualification' in the sense that because a particular country has opted to install and rely on a significant amount of wind generated electricity, it does not necessarily imply it has been a *wise* choice, or that it is suitable for the United Kingdom – as an example, the energy system in Germany was relatively straightforward, based on large scale conventional power plants, which generally produced electricity at a constant rate. But now the system is being shaken up, as the 'Green' energy transition leads to more and more wind generators and solar arrays feeding into the grid. This new system leads to ever greater fluctuations in power generation, with output changing with every gust of wind and every cloud that crosses the Sun - thus is it surprising that in Germany they are now building more coal-fired power stations that offer constant and reliable output.

It is interesting and significant to note that on Saturday 4th November 2006 Europe experienced a power blackout (outage) which affected countries as far apart as Germany, France, Belgium, Spain, Austria and Italy - over 10 million Europeans were left without power and it was observed that the grid's inability to cope with the ever-increasing amounts of wind generated power might have been the cause - although at a press conference on large-scale integration of wind energy on 7th November 2006, Daniel Dobbeni, the President of the European Transmission System Operator (ETSO) said, "We don't know whether wind is responsible or not." However, when you consider that power grids are currently designed to distribute electricity from centre to periphery i.e. from large power stations out to the customer, there is a fundamental problem with connecting wind farms to the network - all wind farms and domestic mini-wind generator installations (by their very nature), feed power into the network via the opposite route which creates problems for the system. See also 'Problems in Local Power Networks' later in this chapter.

The German power company, E.ON have said the European problems begun in north-western Germany, where its network became overloaded. Now bearing in mind that Germany is the European leader in wind generation it does give food for thought and should we regard this unfortunate incident as a 'window to the future' and a 'gypsies warning' to the headlong dash into the proliferation of wind generation - in fact, after reading this chapter on wind generated electricity, I feel the reader will have absolutely no doubt to the madness of it all. It should be clearly noted that the arguments put forward in this book against the proliferation of wind driven generators in our finite and beautiful countryside, apply equally to other similar geographical, heavy populated centred and climatic environments as found in the United Kingdom.

Having stated the above, it is accepted there are places around the globe where the employment of relatively small wind generators have an arguable case. These are at places such as cattle farms in the Australian out-back, places which make any connection to an electrical grid system practically impossible due to distance and economics; the same criteria applying to cattle ranches in the USA and other such similar areas. Obviously, when there is no wind or indeed too strong a wind (see later), there is no power generated, so it makes sense to have a backup source of electrical energy in the form of a petrol or diesel driven generator. Parts of the United Kingdom that could argue a case for a wind generator would be remote dwellings on islands off the coast of Scotland, where connection to the local distribution network or Grid would not be viable due to geographical placement and economics – but of course this need would have to be weighed in the balance with local environment and wildlife - backup again would be required for a guaranteed power supply.

It should be appreciated that even if the wind generator provided direct current (DC) to charge a bank of batteries, it would still be sensible to have a petrol or diesel driven generator, for prolonged periods of very strong, weak or no winds as batteries have a finite charge.

The complete solution for a remote dwelling would include a direct current (DC) wind generator, a bank of batteries to store the electrical energy, solar (photovoltaic) cells to help charge the batteries when the Sun is shining, and the use of a petrol or diesel driven generator for backup, when there is no useful wind, nor sun. A controller would be necessary to not only ensure the batteries are not over charged, but also to divert power to another useful source such as water heaters or pumps. Alternating current would be obtained from the batteries by employing a suitable 12 volt to 230/240 volt inverter. A system designed to incorporate an alternating current wind generator, would require a rectification unit, converting alternating current to direct current to charge the batteries.

The author recognises that at locations on the planet where there is a steady wind source, and there is no measurable impact on the local environment nor wild life, then it would be churlish to deny large scale wind generation in such areas - but where are there such areas - this is far from the case for the United Kingdom - and for many other parts of the globe. It must be recognised that a large wind generator requires a 'cut-in' wind speed of about 16 kilometres per hour (10 miles per hour) to start generating electricity - they reach peak power output at around 53 kilometres per hour (33 miles per hour) - and at very high winds speeds, greater than 80 kilometres per hour (50 miles per hour) the wind generator has to be shut down, otherwise structural damage will occur.

An electrical generator of any kind works on the principle that a conductor moving through a magnetic field has a current induced into it - that conductor must be moving at a certain speed through the magnetic field to reach a certain current level which, when applied to the load, results in a certain output voltage. It is important to note the speed of an

AC wind generator determines the frequency of the generated alternating current - thus it follows that a wind generator must get up to a certain speed to generate a certain voltage and frequency of operation.

At very low wind speeds, there is insufficient torque exerted by the wind on the wind generator blades to make them rotate. However, as the speed increases, the blades will begin to rotate, turning the drive shaft to the gear box, and then the generator shaft for the generator to produce electrical power. The speed at which the wind generator first starts to rotate and generate power is called the cut-in speed and is typically, as mentioned earlier, about 4 metres per second - or 16 kilometres per hour (10 miles per hour) as mentioned above.

It is very important for the reader to understand that just because the rotor and the blades are spinning (start-up speed), it does not mean that the generator is producing power - the start-up speed is the minimum wind speed needed for the rotor and the blades to begin spinning, this low rotational speed will not provide any usable electric power - at low wind and rotational speeds the wind generator will produce no power until the wind speeds reach the required cut-in speed for that particular wind generator. All wind generators have a distinct start-up speed and a cut-in speed.

As the wind speed rises above the cut-in speed, and hence the rotational speed of the blades, the level of electrical output power rises rapidly. However, at around 53 kilometres per hour (33 miles per hour), the power output reaches the limit that the electrical generator is capable of. This limit to the generator output is called the rated power output and the wind speed at which it is reached is called the rated output wind speed. At higher wind speeds, the design of the wind generator is such to limit the power to this maximum level and there is no further rise in the output power. How this is done varies from design to design but typically with large wind generators, it is achieved by adjusting the blade angles so as to maintain the power at a constant level.

As the speed increases above the rate output wind speed, the forces on the wind generator structure continue to build and, at some point, there will be a risk of damage to the rotor. Therefore, a braking system is employed to bring the rotor to a standstill. This is called the cut-out speed and is usually greater than 80 kilometres per hour (50 miles per hour). The cut-in and cut-out speeds are the operating limits of the wind generator - by keeping within this range, the available energy (useful power) is above the minimum threshold and the structural integrity of the wind generator is maintained.

It cannot be reiterated enough times (so the pro-wind advocates can fully understand) that at low wind speeds of less than 16 kilometres per hour (10 miles per hour) a wind generator does NOT produce any meaningful power, as there is insufficient energy in the wind to productively turn the propeller. Additionally, NO electrical energy is produced when the wind speed is greater than 80 kilometres per hour (50 miles per hour) as the unit has to be shut down to prevent damage occurring. Therefore it makes for good common sense to carry out an assessment of the potential wind energy at the site a wind generator is to be proposed - no one in his right mind would install a wind generator in the Doldrums for the obvious reason of lack of any wind - apart from any other factors. Conversely, it would be silly to install a wind generator where there are continuous and extremely high winds - wind generators, as we have stressed above, have to be shut down when the wind exceeds 80 kilometres per hour. So an ideal site would be where the wind blows constantly at a speed in excess of 16 kilometres per hour and less than 80 kilometres per hour.

Due to its very nature, the wind is not that obliging across the United Kingdom land mass, rising and falling dramatically and difficult to forecast, as we all know only too well. It is prudent to recognise the accuracy of the UK weather forecasts, such as the now infamous October hurricane, which hit the south coast of the United Kingdom during the night of the 15-16 October 1987. How many tornados can we expect similar to the one that hit the centre of Birmingham during 2005, and the one that took the roof off houses (up to 150 houses were affected) in northwest London during 2006?

When choosing a site for a wind generator it is important to recognise that obstructions on or near the ground disrupt the flow of wind – increasing the height will have a marked effect on wind speed and power. One way to calculate the increase in wind speed with height is to employ what is known as the 'power law' equation:

$$V = (H/H_o)^\alpha V_o$$

Where V = wind speed at new height
V_o = wind speed at original height
H = new height
H_o = original height
α = the surface rough exponent

Now the rate at which the speed of the wind increases with height varies with the surrounding vegetation and terrain. For example, the increase is greatest over rough terrain or numerous obstacles such as trees and shrubs, but the lowest is for surfaces such as smooth water or ice - the surface roughness exponent (α) can vary from 0.1 for ice and water to 0.28 for woodlands and suburbs. It is important to fully appreciate that wind velocity (speed) increases with height, as this in turn has a very significant effect on the wind power available – doubling the height of the wind tower increases the power available by 34 per cent and increasing the tower height by five times almost doubles the wind power available – this is because wind power has a cubic relationship with wind velocity.

Number of Blades

Historically, there have been lengthy debates over the merits of using two or three blades for a wind generator propeller. Obviously the two bladed versions are cheaper to manufacture than three, having fewer blades; although the three bladed propellers will run more smoothly and generally will have an increased life span. Thus in the long term the three bladed version is proving to be the better option. The majority of small wind generators use materials such as fibreglass (glass reinforced polyester) for their blades, with a few still using wood, although aluminium is not used due to its tendency to metal fatigue. More recently composites using carbon fibre are being employed instead of glass.

Mini and Micro Wind Generators

Before evaluating large wind generators it will be prudent to look at the sensible deployment of mini and micro wind generators. It is only fair to say that both these type of generators do have a use and are employed gainfully around the world - their output may be very small when compared to large electrical generation, but they do make a significant difference to the lives of people in remote areas, and who would deny the yachtsman or caravan owner their small wind generator for charging batteries and energising electrical equipment. I can offer no reasonable argument (taking into account wildlife and the surrounding landscape) against the use of this type of generator at remote homesteads, farms and ranches - or on remote islands.

If I lived on a remote farm or ranch far from any electrical distribution or grid system, then to satisfy my electrical needs, I would possibly employ an integrated system (as mentioned earlier) supplying a bank of batteries housed in a small building or shed. The output from the batteries would obviously be direct current, or could be an alternating current achieved by use of an inverter, but should the need demand, both types (DC and AC) of supply could be accommodated.

To keep the batteries in a charged state I would employ solar cells complimented by a small or micro wind generator – when the batteries are in a fully charged state, and to stop overcharging, consideration will have to be given to what to do with any excess electrical energy from either the solar cells or wind generator. I would pay careful attention to both the sighting of the solar cells and wind generator, with the latter being mounted on a pole or tower well away from the 'wind turbulence shadow' of any building – I certainly would not be foolish enough to mount it on a building for reasons which we will explore later. For periods when there is unusable wind, poor solar radiation, mechanical failure or a maintenance requirement then a backup diesel or petrol generator would be needed and employed - the capacity of the batteries, solar cells, wind generator and fossil fuelled generator would be determined by household and farm/ranch needs.

It is important to clearly understand that the mounting of small wind generators on a roof of a building is a complete waste of time, especially in urban areas. To even think of mounting a wind generator on a chimney stack or at the gable end of a house is nothing short of madness – older houses can be subjected to serious structural damage from the powerful sideways forces incurred as the wind pushes against the wind generator.

Remember it is of paramount importance to consider the integrity of the building so as to avoid any structural damage and the services of a competent surveyor will be a must - be fully aware that you are mounting a potentially dangerous and dynamic piece of machinery to your property – it cannot in any way be compared to the static and relatively light Yagi type of television aerial or indeed that of a satellite dish.

Due to the roofs, chimneys and other obstacles, the turbulence of winds in cities and towns, forces the wind generator to swing around constantly to 'hunt' for the wind causing the blades to keep changing speed. Next there is the noise consideration. I had a wind vane mounted on the gable end of my detached garage and at times it became quite noisy under conditions of extreme wind turbulence. When inside the garage I was very conscious of the roof structure amplifying the vibration - so I dread to think of the noise that could be created by a wind generator - then there is the external noise and visual aspect - will it upset the neighbours?

It is no secret to informed people that the majority of small wind generators are noisy when governing in high winds, and as with everything else, some are worse than others. Lastly, but most importantly, is how effective will it be in producing electrical energy – will you be wasting your time and money?

The answer to this question is a definite YES to those mounted on buildings in urban areas. Any electrical engineer worth his salt, or indeed, most reputable manufacturers of small wind generators will tell you not to bother as it is really not worth the trouble. But should you foolishly insist on installing a small wind generator for home use then you will be more than wise to consider mounting the device on a pole or small tower well away from any buildings – quite impracticable I would suggest in an urban environment.

When considering the sighting of the pole or tower it needs, according to some authorities, to be placed at a distance of at least twice the height of the building on the prevailing wind side of the building, and at least at a distance of twenty times the height of the building on the opposite side of the house to avoid power robbing wind turbulence – does your urban house fit this bill – I doubt it very much.

As with solar panels it pays to be sensible and knowledgeable – it would certainly be daft to place solar panels on a north facing roof. Regarding wind flow a reasonably good method of determining the flow at a proposed site is to simply to fly a kite – by attaching streamers to the kite line and observing their motion will give a good indication to air flow – those streamers near the ground will tend to roll and flap, whilst those further up will fly more smoothly. It is in this area the wind generator should be located above the ground.

Experience has shown that the minimum height above ground should be of the order of 10 metres (33 feet), although most sensibly installed wind generators are on towers much taller, so as to take advantage of the stronger and less turbulent wind flow higher up.

To conclude, considering the required distance away from any obstruction and the height necessary for effective electrical energy generation obviously makes the installation of wind generators in an urban environment far from an attractive proposition – and no doubt the reader will have better things to do with their money.

Large Wind Generators

Generating electricity from the wind is not a new concept as the technology goes back to the winter of 1887-88 when the American, Charles F. Brush (1849-1929) built the first automatically operational wind generator, in Cleveland, Ohio, USA; his generator was known as The Giant Brush Windmill and utilised a 12 kW direct current generator. It was a very large construction consisting of a large disc shaped rotor which had a diameter of 17 metres (50 feet), which incorporated 144 rotor blades made of cedar wood. The modern wind driven machines the Wind Industry employ are obviously of better design using the latest technology, materials and computer control techniques. But, remember, they are basically no more than an electrical alternating current generator driven by a shaft connected to a propeller, which is wind driven; the larger wind generators necessitating a clutch and gearbox between the propeller shaft and the generator.

As an engineer I have never been comfortable with the term Wind Turbine as I contend the term is purely a clever marketing ploy which conjures up a much more sophisticated and powerful machine than it actually is. To be sure, a turbine is a fairly complex piece of machinery consisting of numerous blades along a shaft fitted within a casing i.e. an engine in which steam, water or gas is made to spin a rotating shaft by pushing on angled blades, like a fan.

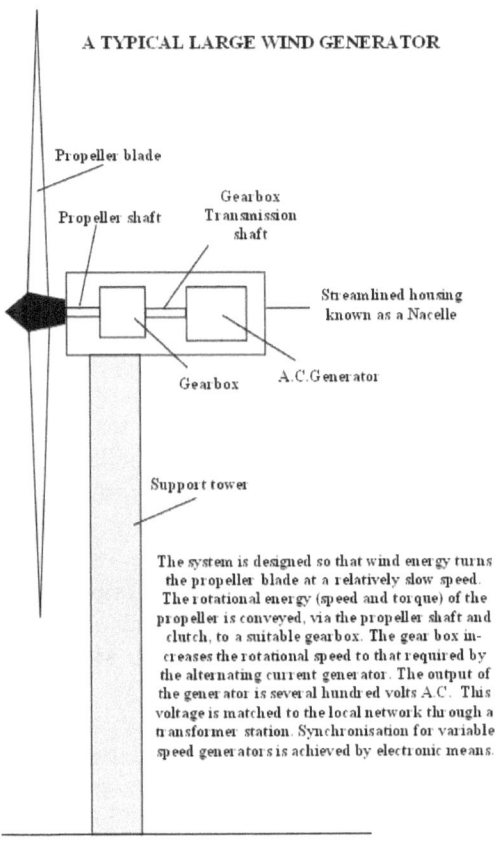

A TYPICAL LARGE WIND GENERATOR

The system is designed so that wind energy turns the propeller blade at a relatively slow speed. The rotational energy (speed and torque) of the propeller is conveyed, via the propeller shaft and clutch, to a suitable gearbox. The gear box increases the rotational speed to that required by the alternating current generator. The output of the generator is several hundred volts A.C. This voltage is matched to the local network through a transformer station. Synchronisation for variable speed generators is achieved by electronic means.

Turbines being among the most powerful machines with steam turbines driving generators in power stations and ship's propellers, with water turbines spinning the generators in hydroelectric power stations, gas turbines powering jet aircraft et cetera. The wind driven machines the Wind Industry employ are not turbines as defined above, but simply, electrical generators connected directly, or via a gear box and clutch, to a large wind driven propeller - go out and have a look at one of these 'monsters' and judge for yourself – the three bladed versions could be compared to a scaled up version of a Lancaster bomber propeller.

The propeller is an accurate description as it does indeed propel or drive the 'shaft' which drives the electrical generator, such as in the same context as the propelling 'shaft' between the gearbox and the rear axle in a 'motor vehicle'. Therefore these propeller blades are not a turbine in the same sense that the sails or vanes of a conventional windmill are obviously not a turbine.

The truth is laid bare if we look at turboprop aircraft which employ a compressor and turbine to drive a propeller - there are two main parts to a turboprop propulsion system, the core engine and the propeller. The core is very similar to a basic turbojet except that instead of expanding all the hot exhaust through the nozzle to produce thrust, most of the energy of the exhaust is used to turn the turbine.

The drive shaft is connected via a reduction gear box to a propeller that produces most of the thrust. The exhaust velocity of a turboprop is low and contributes little thrust because most of the energy of the core exhaust has gone into turning the drive shaft – thus a turboprop engine can be said to be simply a jet engine attached to a propeller - the turbine at the back is turned by the hot gases, and this turns a shaft via a reduction gear to turn the propeller – a number of small airliners and transport aircraft are powered by turboprops.

Modern turboprop engines are equipped with propellers that have a smaller diameter but a larger number of blades for efficient operation at much higher flight speeds - to accommodate the higher flight speeds; the blades are scimitar-shaped with swept-back leading edges at the blade tips.

Another good comparison of incorrectly calling a wind driven generator a wind turbine would be that of the helicopter and their large blades which are known as rotor blades – nobody in their right mind would refer to helicopter rotor blades as turbines!

Thus returning to our wind driven generator the propeller by itself is not a turbine, the shaft from the propeller to the gearbox is obviously not a turbine, and finally the shaft from the gearbox to the generator is not a turbine, nor indeed, is the electrical generator.

The streamlining of the nacelle does not constitute a turbine. Indeed, the whole arrangement is not a turbine. A turbine, as has been pointed out, is a powerful machine and is basically a set of blades fixed on a shaft which rotates inside a casing - so how on earth can these wind machines be defined as turbines - Charles Parsons, the inventor of the steam turbine, must be turning in his grave (pun intended).

Photograph of 25 kW wind generator on display clearing showing, hub and part of propeller, slow speed shaft leading to gearbox, high-speed shaft leading to generator, with nacelle cover standing on ground.

When people first harnessed the wind to drive a mill they called the whole device a windmill (wind driven mill); when the wind was used to pump water the device was called a wind pump (wind driven pump)!

Therefore it follows, if only by convention, that using the wind to drive an alternating current generator, the whole device should be called a wind generator, that is, a wind driven generator. The use of the word turbine is totally misleading - how can a very large propeller driving an alternating current generator possibly be correctly defined in engineering terms as a wind turbine - I would suggest the most accurate term is a WIND DRIVEN GENERATOR (WDG), or WIND GENERATOR (WG) for short, it is as simple as that - there is no ambiguity or any pretension to other than what it is.

The language of the Wind Industry can be very misleading insomuch that the Wind Industry will talk in the same breath of wind farms and conventional power stations. Now these are two entirely different 'animals' when it comes to the production (magnitude) of electrical energy and security of supply. By talking about them in the same breath, effectively bestows on the wind farm a higher status than it deserves - as the same for the term wind turbine – something better than what it is – it is the Wind Power Industry wishing to obfuscate and confuse the public by encouraging the use of ambiguous terms – for readers old enough it is useful to remember the old Saturday afternoon matinee with their Western films and the 'injuns' complaining that, "White-man speak with forked tongue!"

Load factor

A wind generator is not, by its very nature, a very effective provider of continuous electrical energy, and one method of determining its overall effectiveness is to compare the generator's actual output (energy) over a given period with the amount of energy the generator would have produced if it had run at full capacity for the same period of time.

This method of determination, as we have seen in Chapter Two, is known as calculating its Load Factor.

Where Load Factor = Actual amount of electricity produced over time/Electricity that would have been produced if generator operated at maximum output 100 per cent of the same time.

The reader may come across the term, Capacity Factor, but please note it is the American electrical engineering professions synonym for Load Factor and this term is not recognised by British Standards Institution; although it is used by the British Wind Energy Association (BWEA) for reasons best known to them.

The Load Factor of a wind generator in the UK is typically of the order of 29 per cent or less, and although a modern wind generator produces electricity 70-85 per cent of the time, it generates different outputs depending on the wind speed - thus it should be clearly understood that for most of the time the electricity produced is not 'useful electricity' - over the course of a year, the wind generator will typically generate useful electricity for about 29 per cent of the theoretical maximum output, and to reiterate, this is known as its load factor.

Another way to look at what is meant by 'useful electricity' is to consider the case of a fully charged car battery that becomes, for whatever reason, almost flat – a situation a lot of us have no doubt found ourselves in. A car battery suddenly found in this condition will not have sufficient energy to operate the car starter motor and thus start the car engine, but the battery may have sufficient energy for say operating the side lights and maybe the car radio. Therefore in both cases we can say the battery produces electricity, but this is misleading as a good battery will start the car, but the flat battery will not, although being capable of operating side lights and the car radio – in both cases we can say the battery produces electricity, but only in one case is the electricity fully productive - similar to the claim of wind generators producing electricity for 70-85 per cent of the time – this is why the term Load Factor is more meaningful and honest.

A typical fossil fuelled power station would have a Load Factor on average of 50 per cent - recognising conventional power stations do have their load factors constrained by human intervention for various operational/supply reasons. Thus by comparison a wind generator/wind farm is not very impressive and the only good thing is that the wind is free - when it is blowing - and then, it must not be too little, or too much. The British Wind Energy Association quotes load factors of about 30 per cent for wind farms and an average of 50 per cent for a conventional power station.

The old Department of Trade & Industry (DTI), DUKES 2005 Table 5.10 gave average UK-wide load factors for 2004 as follows:

Combined Cycle Gas Turbine (CCGT) 60.3 per cent.
Coal fired 62 per cent.
Nuclear 71 per cent.

The Department of Energy and Climate Change (DECC), DUKES Table 5.90 Plant loads, demand and efficiency; (www.decc.gov.uk) gave average UK-wide load factors for 2008 as follows:

Combined Cycle Gas Turbine (CCGT) 69.3 per cent.
Coal fired 56.7 per cent.
Nuclear 49.4 per cent.

And for 2009:

Combined Cycle Gas Turbine (CCGT) 64.2 per cent.
Coal fired 38.5 per cent.
Nuclear 65.6 per cent.

Regarding wind farms data released from Ofgem indicated a load factor of only 23.8 per cent for the summer-autumn 2005 period for the Cefn Croes wind farm near Aberystwyth, Ceredigion, which came online during June, 2005. This wind farm in terms of capacity (58.5 MW) is, at the time of writing, the largest onshore wind farm in Wales; other sites in Wales such as Llandinam and Carno have produced a load factor figure of only 22 per cent. Cefn Croes with its 39 large wind generators, covers an area of 7.5 square kilometres of what was once beautiful Welsh countryside - and what of the damage caused to the landscape by its construction including the felling of some 80,000 trees (the lungs of Wales) - what a mindless act of blatant vandalism and desecration of wonderful countryside – and for what – peanuts!

The Department of Energy and Climate Change (DECC), Load Factors for Electricity generated from renewable sources, DUKES 7.5, had the following information:

Type	2004	2005	2006	2007	2008	Average
Onshore	26.6	26.4	27.2	27.5	27.0	26.94
Offshore	24.2	27.2	28.7	25.6	30.4	27.22

The Department of Energy and Climate Change (DECC), Load Factors for renewable electricity generation, now DUKES 6.5, had the following information at January, 2015:

Type	2009	2010	2011	2012	2013	Average
Onshore	26.9	23.7	29.8	29.2	32.3	28.38
Offshore	27.2	21.7	27.2	26.2	28.9	26.24

Note: Average figures derived by author.

The figures for the last five years show a load factor of less than 29 per cent for both Onshore and Offshore wind generators. Indeed the average figure for the last five years for onshore is only 28.38 and for offshore 26.24 – not an impressive record.

The Department of Energy and Climate Change (DECC), The Capacity of, and electrical generated from renewable sources, Dukes 6.4, had the following information at January, 2015:

All figures: MW

Type	2009	2010	2011	2012	2013
Onshore	3468	4055	4620	5899	7513
Offshore	951	1341	1838	2995	3696

The actual electricity generated in GWh:

Type	2009	2010	2011	2012	2013
Onshore	7529	7136	10,347	12,112	16,992
Offshore	1754	3044	5126	7549	11,441

Electricity: Chapter 5, Digest of United Kingdom Energy Statistics (DUKES) states the total electricity for the UK was 373,581 GWh (374 TWh) for 2013, which puts the total UK wind generation of 28,433 GWh (28 TWh) into perspective.

It is not rocket science to conclude that wind generators have serious limitations in producing electrical energy, being at the mercy of the vagaries of the wind and thus operating at less than 29 per cent of their total capacity; these devices have other short comings which will be examined later in this chapter, but before doing so, we will see what is meant by a wind farm.

Wind Farms

This is another misleading and irritating use of words - the word farm identifies with something ecologically friendly and productive, and therefore by association, wind farm also sounds very green and productive! But remember the construction and installation of a wind generator itself produces significant amounts of CO_2 and they do have a very significant impact on the surrounding landscape – how can they not, unless they are invisible, The wording also implies that a wind farm is a meaningful and continuous producer (of electrical energy) – this is not so, as the DECC statistics clearly demonstrate.

A wind farm is no more than a large grouping of ineffective wind generators contained on a single site - the 'environmentally friendly tag' associated with wind generators and which the Wind Industry would have us believe in, is greatly exaggerated. According to British Nuclear Fuels (BNFL) a wind farm

A wind farm is no more than a large grouping of wind generators contained on a single site as shown above.

covering an area of 712.25 square kilometres (275 square miles) would be required to produce the equivalent electrical output of Wylfa nuclear power station (now decommissioned) - this nuclear station had an installed capacity of 980 MW of electricity and a footprint of less than 150 acres. Remember 1 square mile is equal to 640 acres, so we are comparing the industrialisation of 176,000 acres to a mere 150 acres – to even consider a wind farm of this size would be an act of mindless folly. It is also very important to note the average electricity output of this 'huge' wind farm would merely match the output of the nuclear power station – that is when the wind is blowing at the correct speed - it would not result in the closure of any fossil-fuelled or nuclear plant - unpredictable wind power has to be 'shadowed' at all times by a secure, controllable source of electricity such as fossil fuelled or nuclear powered stations. So much for cutting down on CO_2 emissions or reducing nuclear energy - a bit of a fraud, I think you will agree.

Backup

Due to the very nature of the wind in the UK a wind generator cannot be relied upon to produce a continuous, uninterrupted supply of electrical energy - under conditions such as high pressure in winter or summer, a wind generator or indeed a wind farm, may not produce any meaningful power at all - then at other times the vagaries of the wind have to be considered as at any one moment the wind may be blowing at a useful speed, then the next moment, just a breeze. Therefore it follows that some form of backup that can react to all these conditions will be required if security of supply is not to be in danger.

The difficulty with backup for wind farms is the question of how much backup is to be kept running at any one moment in time – running total backup by a conventional power station would be rather silly as then there would be no need for the power from the wind farm and as such it would be wasted. So the question is how much spinning reserve would be required that would be economically viable to satisfy a variable demand for power?

Without going into too much detail spinning reserve is generation capacity that is on-line but unloaded and that can respond to compensate for generation or transmission outages within 10 minutes – a 200 MW power plant running at say 90 MW may be considered part of the spinning reserve – the spinning reserve in this example being 90 MW which is known as the reserve capacity.

Non-spinning reserve (supplement reserve) refers to power sources which are presently not on-line, but can brought into the system if the demand rises - the non-spinning reserve should be able available to supply the load within 10 minutes.

The tricky part for the power producing industry is how to determine the total amount of backup for wind generation that is considered economically viable - who can say for sure, where and when the wind will blow, and obviously the greater the number of wind farms, spread over a greater area, the bigger the problem.

If a single wind farm is connected intermittently to the National Grid it would not pose too much of a problem (just synchronisation) as the Grid has approximately a 20 per cent margin over peak demand – this being required to guard against line transmission and/or generator failure. The real problems start to arise when more and more onshore and offshore wind farms are connected to the Grid as apart from synchronisation, there will come a point when serious consideration has to be given to providing additional resources, with the ability of being able to assess correctly the capacity and number of stations necessary to ensure system security of supply.

It is important to recognise that you cannot just switch on steam driven turbines similar to the action of throwing a light switch and having immediate power. You have to activate a heat source at the power station to convert water into steam to drive the turbines. Coal-fired stations will take several hours from cold before they can produce any useful energy – gas, oil and nuclear stations still have to boil the water for steam, although they can be brought on-line quicker than coal.

One of the strangest (and perverse) arguments put forward by proponents of wind driven electrical energy in the UK is the claimed *significant* saving in carbon emissions. How can this be, when as we have seen, wind generators for a guaranteed continuance of supply, require backup by conventional power stations.

The vacuous argument of UK wind power supporters is put into even more perspective when the pollution from the industrialisation of China, India and Brazil are brought into the equation - noting that China is the planet's top polluter! For example, do these supporters of wind power not realise that during 2005 China set out on a seven-year programme to build over 500 new coal-fired power stations – do they not know that most of China's electricity is generated by burning dirty coal in dated power stations creating horrendous air pollution – nearly half a million Chinese are killed by smog each year – perhaps they do not realise it is predicted that in about ten years' time there will be 140 million vehicles on China's roads - are they not aware of the massive building programme going on in China demanding millions of tons of concrete and other associated materials - imagine what that will do for global CO_2 emissions.

But then what about America, currently the second greatest contributor to CO_2 gasses on the planet, and then India - bearing in mind that of the global total of 27 billion tons of carbon dioxide put into the air each year, the UK accounts for approximately 550 million tons – which is 2 per cent of the global total – so these pro-wind people really need to get their facts straight and put things into perspective – no doubt they will go into denial when confronted with the truth.

It is important though to discriminate between CO_2 emissions and pollution such as carbon monoxide and very small particulates, for CO_2 is necessary for all life on Earth and a reasonable additional amount to the atmosphere can be beneficial and will manifest itself by increased plant growth – that is why the Dutch pump CO_2 into their greenhouses – not to poison the plants inside, but to enhance healthy plant growth.

Therefore, and not diminishing the pollution from industry, it is very important not to be blinded by other truths and significant factors that pose a greater threat to the wellbeing of us all, and are the Greens purposefully missing the elephant in the room – that of land transport - motor vehicle pollution, and in particular, that of the diesel engine. The threat to health is already with us when a high pressure system dominates over the UK and the air flow is from Europe, coupled with dust blowing up from storms in the Sahara. This combination resulted in extreme air pollution (smog) over London and other parts of the country on Friday 10th April, 2015 with the temperature reaching 21.9 ^{0}C at St James's Park in London and 21.3 ^{0}C at Heathrow Airport. People with lung and heart conditions were warned against exercising outside. Make no mistake as unless this situation is addressed purposefully it will result in an ever increasing debilitating scenario with normally healthy people beginning to succumb to breathing difficulties.

So why are the Greens concentrating so much on power stations and ineffective wind farms when the greater threat to the planet are the millions of vehicles, with their exhaust fumes, operating around the planet – not forgetting the tiny particles of carbon found in air that has been polluted by traffic and known as PM2.5s. These are particles less than 2.5 microns wide - a micron is a millionth of a metre. These particulates can bypass the mucus in our airways that trap dust and pollutants so rendering this 'filtering system' ineffective and allowing pollutants to enter our bodies. This is particularly dangerous for those suffering from heart and lung problems - but what of the rest of us - do we think we are immune…?

Unfortunately and as mentioned above, the answer is a big NO - we are not immune! According to the Daily Mail (Dec, 2014), and as an example, scientists argue and warn that the air pollution levels are so high on Oxford Street, London, that just by spending two hours there can cause significant stiffening of the arteries. Indeed just recently a group of MPs called for planning guidelines to be changed so schools and care homes could no longer be built near pollution hotspots; according to a study at St George's hospital, South London, one in fifty heart attacks that lead to admissions at London hospitals may be triggered by air pollution.

It would appear the lessons of the last two centuries are soon forgotten when Londoners suffered from extreme 'smog' – this is a type of air pollution consisting of smoke and fog (commonly known then as pea soup fog) – and was a familiar and serious problem for Londoner's from the 19th century to the mid-20th century, due to the burning of large amounts of coal in homes and factories. Smog consists of soot particulates from smoke, sulphur dioxide and other components. Smog itself is simply airborne pollution which may obscure vision and cause various health conditions. It is caused by small particles of material which become concentrated in the air for a variety of reasons. Commonly, smog is caused by an inversion, in which cool air presses down on a column of warm air, forcing the air to remain stationary.

Modern smog (Photochemical smog) is a type of air pollution derived as a result of vehicular emission from internal combustion engines and industrial fumes that react in the atmosphere with sunlight to form secondary pollutants that also combine with the primary emissions to form photochemical smog. The atmospheric pollution levels of cities such as Mexico City are increased by inversion that traps pollution close to the ground - it is usually highly toxic to humans and can cause severe sickness, shortened life or death.

Photochemical smog is a unique type of air pollution which is caused by reactions between sunlight and pollutants like hydrocarbons and nitrogen dioxide. Although photochemical smog is often invisible, it can be extremely harmful, leading to irritations of the respiratory tract and eyes. In regions of the world with high concentrations of photochemical smog, elevated rates of death and respiratory illnesses have been observed. Some of the particulate matter in the air can oxidize very readily when exposed to the UV spectrum. It doesn't have to be that sunny for photochemical smog to form as UV light can penetrate clouds. The pollutants released through human activity in this situation are known as 'primary pollutants' and they include sulphur dioxide, carbon monoxide, and other volatile organic compounds. When these compounds interact with the Sun, they form 'secondary pollutants' like ozone and additional hydrocarbons.

The number of vehicles on our roads are steadily increasing on a global scale, just think of all the vehicles you encounter daily on the roads of the UK that are all burning oxygen and spurting out obnoxious pollutants, then contemplate all the vehicles on the roads in France, Germany, Italy, Spain, Africa, India, Australia and the North and South American continents – not a day or night passes without tons of pollutants pouring into the atmosphere.

But then what of the dust particles created from tyre, clutch and break lining wear - surely this wear and tear breaks down into fine dust as it certainly does not just disappear into thin air - then what of road surface wear and tear, again the same forces are surely at work breaking the material down into a fine dust.

Moving away from land transport it would be remiss of me to ignore another 'elephant in the room' and that of global commercial and military aircraft.

According to 'Environmental Protection UK' which is a national charity that provides expert policy analysis and advice on air quality, land quality, waste and noise and their effects on people and communities in terms of a wide range of issues including public health, planning, transport, energy and climate, in 2011, approximately 200 million passengers passed through mainland UK airports. This was a return to growth, following a recent period of decline in passenger numbers and air transport movements between 2007 and 2010.

Government forecasts predict that this will rise to 255 million in 2020 and 313 million in 2030. As with the internal combustion engine, aircraft are responsible for an increasing proportion of air pollutant emissions, both at local and global level. Aircraft engines generally combust fuel efficiently, and jet exhausts have very low smoke emissions. However, pollutant emissions from aircraft at ground level are increasing with aircraft movements. In addition, a large amount of air pollution around airports is also generated by surface traffic. The main pollutant of concern around airports is nitrogen dioxide (N_2O). N_2O is formed by nitrogen oxide (NOx) emissions from surface traffic, aircraft and airport operations. PM2.5 is also of concern, since particulate emissions from jet exhausts are almost all in this fine fraction. NOx in the lower atmosphere contributes to the production of ozone; ozone in the lower atmosphere is a pollutant, and contributes to global warming. Nitrogen oxides from high-altitude supersonic aircraft are thought to damage the stratospheric ozone layer, the protective layer that filters out harmful radiation from the Sun.

I would suggest that CO_2 emissions from aircraft are not a significant problem as they contribute about 3 per cent of the global total - the more worrying problem being water vapour aircraft emit, which shows up as condensation trails (contrails) behind high flying aeroplanes. Indeed a few years ago this was amply illustrated when sunbathing during a very sunny morning in June on a beach in Tenby, Wales; it was not long before the sky was covered in a tenuous high cloud due to various aircraft contrails converging. Unfortunately this had the effect of diminishing the heat from the Sun, putting an end to a glorious sunny June morning and thus making sunbathing quite uncomfortable – boy did my wife and I curse those aircraft at that time, whilst agreeing the health of our planet was apparently not in safe and capable hands in spite of all our modern knowledge and advanced technology.

I wonder just how many people are aware that water vapour (contrails) released at high altitude of greater than 30,000 feet has a bad effect on the atmosphere many times greater than it would if released in the lower atmosphere, as the vapour does not condense into clouds and rain in quite the same way, and as such collects more dust et cetera - this effects the way the atmosphere behaves at those heights – could recent weather anomalies be attributed to some degree to this effect, and if so, will the situation worsen with more aircraft filling our skies?

All rational thinking people are totally in favour of a clean and pollution free environment and accept there needs to be an intelligent and balanced approach, and it will do well for wind farm supporters to question the cost of not only the construction of, and the running of a wind farm, but also the provisioning of a site with its concrete foundations, access roads, transformer buildings and wires/cables/poles for connecting to the electricity network. The construction and erecting of wind generators require thousands of tons of aggregate and concrete for their foundations, not to mention the material for the approach roads required for the ongoing repair and maintenance of the wind generator. Materials such as cement consume large quantities of carbon-based fuels, such as coal, gas or oil; emitting large quantities of greenhouse gases when burnt in cement kilns during the manufacturing stage. Additionally the aggregate required in concrete making also demands carbon-based fuels for the machinery used in extraction/quarrying process.

This is not to overlook the necessary transportation of the material. All these procedures contribute to the considerable amounts of atmospheric emissions such as carbon dioxide; there is the erection of unsightly poles and/or possibly pylons for connecting to the local distribution or grid network and all what that entails - so in this respect wind farms are not that environmentally friendly that some people would argue, but quite the opposite.

Then there is the inexcusable industrialisation and desecration of the local countryside to consider with its consequential threat to tourism. For example, large numbers of wind generators across the Welsh hills will be devastating for the countryside and the people who live and earn their living there - tourism is the largest rural industry, and earns over £2 billion a year for Wales - as such it contributes 7 per cent to Welsh GDP and far outweighs agriculture, which contributes less than 2 per cent of GDP. It is by far the most important rural earner.

It begs the question as to who would want to visit Wales if the landscape becomes 'industrialised' with hundreds and hundreds of wind generators - the special qualities of the Pembrokeshire landscape, for example, are enjoyed not only by local residents, but by the many tourists that visit the area to enjoy the remote and tranquil areas that are characteristic of the Pembrokeshire countryside – wind generators create an unacceptable adverse effect to the character and visual qualities of this special landscape area – indeed, allowing generator numbers to increase creates an adverse cumulative visual impact on the surrounding area and that of the Pembrokeshire Coast National Park. Surely all Planning Offices in the UK should automatically dismiss any application for a wind farm or a single wind generator unless there are extremely good mitigating reasons to allow an application, and then, only with the agreement of the local populace.

Welsh Government supporters of wind technology claim wind farms will minimise the carbon footprint of Wales - but then why are they allowing the felling thousands of trees (the lungs of Wales) only to be replaced by an industrialised landscape with ineffective wind generators – it begs the question, "Do the authorities know what they are doing?" The same applies to Scotland and its wonderful wilderness and irreplaceable scenery - the magic of the Highlands would be truly lost - it will be a tragedy beyond measure.

In England such places as the Cotswolds and the Mendips would lose their beauty and attraction if wind farms are allowed. Can you imagine how the White Cliffs of Dover would look with line upon line of wind generators upon them - picture these monstrosities across the Yorkshire dales, the Peak District and what of beautiful Northumberland, not to forget the Lake District and areas such as East Anglia – not forgetting Bodmin Moor, Dartmoor, Exmoor - the list goes on and on - to put it in a nutshell, THERE WILL BE SO MUCH LOST, FOR SO LITTLE GAINED - it can be described as the pursuance of a policy dreamt up by lunatics and disturbingly summed up by the Latin, q*uo Deus vult perdere prius dementat* - whom God wishes to destroy, he first makes mad.

Threat to Wildlife

It is enigmatic and somewhat alarming that none other than the Royal Society for the Protection of Birds (RSPB) is in support of wind power in the UK. The RSPB publicised policy is to support wind power, and make sure wind farms are 'carefully sited' – now surely this is stretching credibility a bit too far, as you have to wonder as to what in reality does 'carefully sited' mean? Wild birds are not restricted by any 'sky fence or netting', accepting that certain species of bird are confined, by their very nature, to various geographical areas - you would expect to find Golden Eagles in Scotland, but not flying freely in the centre of London – and yet, unbelievably they are building many wind generators in Scotland. The reader should be aware that for many years critics, such as Mark Duchamp (save.the.eagles@gmail.com), have been trying to warn the public that a 'conflict of interest' has inverted priorities within the charity – but such is the 'prestige' of the RSPB the media appear (prefer) to continue to ignore it.

The general public should indeed be awakened to what is happening here in the UK. Evidence is mounting that a wind generator is a perfect bird trap – whilst appearing to move slowly, wind generator blades travel at speeds of 150 to 300 km/h at the tip, depending on wind speed - they also travel in an orbit – thus a bird about to fly through them may see one blade moving away, but will not always see the next one coming – adverse conditions such as wind, fog, or darkness add to the problem. It is easy to see why the pro-wind farm policy of the RSPB is vital to the controversial wind industry. For without it, the general public might not warm to an industry that is erecting monstrous machines, that desecrate the landscape and are capable of killing 200,000 to 500,000 birds a year (figures from www.iberica2000.org), with many belonging to rare species - especially bearing in mind, and to reiterate, wind farms unquestionably produce unreliable quantities of intermittent, erratic and difficult to control electricity.

What are we to make of the RSPB claim that very significant bird kills are confined at two foreign locations, namely, Altamont Pass (California) and Tarifa (Spain) – these being the exception due to badly sited wind farms! But these two wind farms are NOT badly sited from a wind generation point of view for that is where the useful wind is, and is why the wind industry put the generators there in the first place – so what is the RSPB trying to say?

The Isle of Skye is a natural jewel and a haven for bird life. If it were not for the young eagles that are fledged here and in other Hebridean islands each year, mainland Scotland would be unable to maintain its eagle population. Edinbane and Ben Aketil are two wind farm projects targeting this island - what is worse are the hills selected by the developers are visited daily by young eagles. These may come from anywhere in Scotland, and even England, as young eagles roam the country for years before settling down.

The Altamont Pass wind farm in California (as mentioned earlier) is also located in such a 'dispersion area' for young eagles, and its numerous wind generators killed 2,300 eagles over 20 years. Ornithologists Dr S. Smallwood and C. Thelander, who performed the most complete studies of this infamous wind farm, concluded that the deaths were not imputable to the lattice-type, old model wind generators, as tubular ones killed even more birds – this is logical as they are bigger and sweep larger areas. They also warned that any wind farm erected where raptors fly would be deadly to them. There was nothing particular to Altamont except for the large number of generators and the abundance of raptors. Besides, it was found that eagles and other raptors were attracted by the wind generators.

Thus with this scientifically gathered evidence, corroborated by the news of eagle strikes at wind farms elsewhere in the world, it was obvious that these two projects located in an eagle dispersion area on the Isle of Skye would act as ecological traps for the Scottish eagles, and become a direct threat to their sustainability. As things now stand, golden eagles in Scotland are in 'demographic difficulty'. Any additional mortality will send their population into decline, and additional mortality repeated year after year for several decades may cause the extinction of the species in the UK. Indeed, at the time of writing, there are only 443 pairs left!

So it is puzzling why the RSPB did not strongly oppose this ill-chosen site as they did for the Isle of Lewis – instead they were content with sending routine objections to the decision makers – and of course that failed to convince them. It was reported by Mark Duchamp that whilst he was investigating the Edinbane project, he came upon a letter that was not supposed to be made public, but which was obtained under the Freedom of Information Act. It was sent on April 25th, 2002 by Dr. Alison Maclennan, RSPB's Senior Conservation Officer, Skye & Western Isles, to Simon Fraser, Area Planning Manager, Highland Council, Isle of Skye. The subject being: the proposed AMEC wind farm at Edinbane. On page 4, under, 'Wader and red grouse interest', it read, "They (red grouse) have been known to collide with generator structure and have shown population decline associated with wind farm developments elsewhere."

Now this is surely a contradiction as a senior RSPB field officer had admitted on paper that wind farms had had a significant impact on red grouse populations – the RSPB had stated that wind farms were only a problem for birds abroad, specifically at Altamont and Tarifa as mentioned earlier – but there are no grouse to be killed at these two locations – therefore it can only be concluded that the RSPB has been covering up the mortality of red grouse.

The British public is being misinformed regarding bird mortality at wind farms; denounce Save the Eagles International (STEI) and the World Council for Nature (WCFN). It is contrary to fact to pretend that these industrial structures are 'carefully sited' so as to avoid risks to birds and bats. It is equally false to allege that grouse and other ground-nesting birds don't mind laying their eggs under wind generators, or that raptors avoid these dangerous areas.

In a recent article, The Guardian newspaper states: "Studies in the UK had found evidence that birds of prey in particular avoided wind farms." But if you look closely at the picture shown in the article, you'll notice that the two birds flying between the wind generators are raptors, red kites in fact, which were reintroduced in the UK at great cost. "So! – they avoid wind farms, eh?" – quips STEI's President Mark Duchamp.

In Germany, where a few wind farms have been loosely monitored for bird and bat mortality, the government has disclosed the number of carcasses reported so far: 69 eagles, 186 kites, 192 buzzards, 13 harriers, 59 falcons, 12 hawks, 7 ospreys, plus hundreds more birds of all sizes and even more bats. "These figures are just a small sample of the ongoing massacre," comments Duchamp, who cites this example: "Ubbo Mammen, an ornithologist commissioned by the German government, estimates that 200-300 Red Kites are being killed yearly by wind turbines in Germany." These machines are driving many rare species into extinction, warns Mark.

In the UK, few raptor deaths leaked through what STEI calls "the wind farm cover-up": three red kites, one osprey, and one sea eagle. "Officially, the eagle died of a heart attack," mocks Duchamp. "In the UK, wind farms are not being monitored for bird mortality: this is how the issue is being kept from the public's eye. Scavengers and wind farm employees dispose of the dead bodies, so it is extremely rare for a dead eagle or osprey to be found by some nosy trespasser."

Birds and bats are being slaughtered by the million in other countries. In Spain, the ornithological society SEO/Birdlife recently estimated that the 800 Spanish wind farms were killing between 6 and 18 million birds and bats a year. Unlike birds killed by cars and cats, these include eagles and many other rare species.

But in the UK, bird charities hold the wind industry in great esteem, on account of global warming but also for their financial contributions to bird research, notes STEI. Hence the new study by researchers from the RSPB and BTO, which were just hailed by The Guardian in these terms: 'Wind farms do not cause long-term damage to bird populations, study finds'.

"But raptors have been excluded from the study," remarks Duchamp. "As for the few bird species that were considered, the research is anything but convincing; besides, other studies have shown opposite results." Mark remembers that, years ago, an RSPB officer wrote the following about the Edinbane project: "They (red grouse) have been known to collide with turbine structures and have shown population declines associated with wind farm developments elsewhere."

The BBC, referring to the same study, recently proclaimed: "Wind farms, not major bird mincers." STEI wonders how this conclusion may be drawn from such an inconclusive and suspicious study, whose scope is not mortality, and only targets the 'density' of selected non-raptor species. As for earlier claims that wind farms in the UK are 'carefully sited', Mark notes that many have been placed in the worst possible locations, where they will mince Scottish eagles into extinction: Eishken (aka Eisgein or Eisgen), Pairc, Pentland Road, Edinbane, Ben Aketil, various eagle ranges in Argyll, etc. "Hypocrisy and deceit are rampant," laments Duchamp.

Mark Duchamp is President, Save the Eagles International and Chairman, World Council for Nature.

The following report is just one of many and is reproduced here with the kind permission of Angela Kelly, President of the Country Guardian, and who is rightfully an industrious opponent to the proliferation of wind generation, and I quote:

"The Herald

Watchdog warns wind farm would kill scores of golden eagles

VICKY COLLINS
February 02, 2006

SCORES of golden eagles would be killed by a massive wind farm in the Western Isles, according to a new assessment. Scottish Natural Heritage (SNH) has reviewed Beinn Mhor Power's plans to build 113 turbines on Lewis and now warns that one golden eagle would be killed every three to six weeks by the structures, not every three to six years as originally claimed by the company. A white-tailed eagle would be killed every eight to 15 weeks, rather than every eight to 15 years, according to David MacLennan, Western Isles area manager for SNH. The figures emerged in a letter from Mr MacLennan to the Scottish Executive, obtained under freedom of information laws by an objector to the plan for Eisgein Estate in the south of the island." End of quote.

It is just not birds being slaughtered by wind generators but bats as well, and should the reader wish to avail themselves with more evidence then access the Internet and just put 'bird kills or bat kills by wind generators' into 'Google' or any other search engine and be thoroughly shocked.

Nuisance Factor and Impact on Human Health

What of the very real nuisance value and possible health threat to residents living near wind generators as the following report (with acknowledgement to Angela Kelly, Country Guardian) indicates, and again I quote:

"Cold Northcott wind 'farm', Cornwall.

I've just received a call from Peter Townsend who has the misfortune to live near the Cold Northcott wind 'farm'. He tells me that a piece of blade has broken from one of the turbines and the loose piece thrown about 100 yards. He says that if one of the turbines nearer to the public highway had suffered similar damage, the piece of blade could have landed in the road. The media has been alerted and is sending a photographer. If there's anyone out there who could take some digital pics for circulation that would be most welcome. Peter and Kathleen Townsend objected strongly to the wind 'farm' as soon as it was proposed in the Early Nineties.

They have objected even more fiercely, but as usual, in vain, ever since it started operating in 1993 as they have suffered from the noise and the 'strobing' effect of the turbines. The only 'sympathy' they received from the wind 'farmer' was: - "Well, get some earplugs then" and, to escape the 'strobing' effect: "Move to another room!" End of quote.

Another report from Country Guardian read as follows, and I quote:

"CORNISH GUARDIAN, 11th October 2001.

Wind power so noisy

We, like the schoolchildren in Alistair Wreford's article of September 27, have a constant reminder of the Bears Down Windfarm - the noise. The 16 turbines are between 750 and 1400 metres from our home. When the wind is in a south or south-westerly direction, we are subjected to a disturbing and intrusive deep rhythmic thumping noise. It is a constant irritation during the day, and wakes us and keeps us awake at night. In National Wind Power's publicity and information, and in a personal letter to us, we were assured there would be no noise problem. NWP are monitoring the situation and we can only hope that they will be willing and able to alleviate the situation.

Mr and Mrs P.Lockett, St Ervan, Wadebridge, Cornwall.

(Note: Patrick Lockett is a veterinary surgeon.)"

End of quote.

Apart from birds being slaughtered by wind generators there is the deplorable killing of hundreds of thousands of bats every year simply because they mistake wind generators for tall trees, a study has suggested - published in the Proceedings of the National Academy of Sciences - the National Academy of Sciences (NAS) is a non-profit organization in the United States - researchers were looking to find out why bats interact so closely with the moving blades of wind generators - before the construction of wind generators, instances of bats colliding with man-made structures were rarely observed. However, hundreds of thousands are now killed annually, with most found dead beneath wind generators in late summer and autumn - scientists from NAS found the species most likely involved in these collisions were tree bats - they used video surveillance to study bats flying at night near wind generators for several months.

According to the Bat Conservation Trust (BCT) UK, the discovery of dead bats and birds underneath wind generators in the U.S. and in mainland Europe has led to concerns that research into the siting of these structures is not sufficiently rigorous, and that some have been erected on migration routes of bats and birds.

The siting of wind generators may be an issue for bats in the UK, not only because of the risk of direct collision if generators are placed on migration or commuting routes, but also because of displacement from foraging habitat. The positioning of mid-sized wind generators in hedgerows is also a concern.

The Bat Conservation Trust would like to see monitoring undertaken at existing wind generator sites and monitoring of all new generators, whether large or small. They would also urge that full impact assessments of the potential effect on bats are undertaken, and for post-installation monitoring to be made a planning condition. If the reader wishes to get involved with the conservation of bats then go to: (www.bats.org.uk/pages/wind_turbines.html) and I am sure that BCT will appreciate all the help they can get.

Keeping the Lights On

Having read so far the reader should now be aware that if all our electrical energy supplies were indeed reliant on wind farms, then it is quite obvious the lights would certainly go out when there is no wind, but they would also go out when there is sporadic wind or indeed when the wind is too strong. Thus any country that relied, for total or a significant amount of their electrical energy on the input from wind generators, would have great difficulty in keeping supplies going - attempting to create an effective system would necessitate having to try and forecast accurately for the correct level of wind for a specific geographical location – an almost impossible task in places such as the United Kingdom. It would be essential to plan and provide for adequate backup when the wind fails, and the only feasible solution would be to have a backup source for all of the wind farms, so as to ensure a continuity of supply - anything less would leave the system open to inadequate power provision leading to brownouts and possible blackouts.

According to the energy regulator Ofgem the UK is entering a period of power instability as old coal and nuclear power stations close, especially with their replacements being delayed. The growing proliferation of wind generators and solar parks will not come anywhere near making up this short fall in power generation – such is the government energy strategy – this unbelievable scenario is compounded by government in arranging for 500 MW of diesel (of all fuels) engine backup.

Green Frog Power Ltd, Birmingham, (www.greenfrogpower.co.uk) builds and operates power stations in the UK and say the National Grid needs their power to alleviate stresses on transmission systems when other power stations shut down.

Another UK company that builds diesel generators is YorPower, Yorkshire (www.yorpower.com) which states on their website and I quote, "UK National Grid to Use Standby Generators: Owners of standby generators in the UK could soon benefit from payments from the government to supply electricity to the National Grid.

As reported in the national press the payments could come about as part of the UK Government's plans to spend nearly £110 billion over the next seven years on building tens of thousands more wind turbines. The problem with wind generated electricity, particularly from on-shore wind farms, is that it is very unpredictable with wild fluctuations in the speed of the wind. This unpredictability means that the National Grid has to have other sources of power that it can 'turn on' at a moment's notice to satisfy demand that it can't meet from the wind turbines. The answer National Grid has come up with is to connect thousands of diesel generators, remotely controlled by the grid, to provide almost instantly available back-up for when the wind drops. Owners of diesel generators are being incentivised with fees to make them available to the grid with subsidies equivalent to up to 12 times the rate for conventionally generated electricity and up to £15,000 per megawatt hour (MWh), or 300 times the normal rate of £50 per MWh. Initially, this 'short-term operating reserve' envisaged relying on standby generators owned by public bodies such as hospitals, prisons and military installations which stood to earn hundreds of millions of pounds from the Government. But with the increasing reliance on wind power many private firms are bidding to enrol in the system to earn up to £47,000 a year in 'availability payments' for each MW of capacity." End of quote.

Wind Farm versus a Conventional Power Station

The Wind Industry and their supporters claim wind farms help to prevent the release of CO_2 into the atmosphere, and are an important way in which we can help combat global warming. We have seen earlier that this claim has no grounding - and not one conventional power station has been closed as a direct consequence of a wind farm - nor will there be for the real necessity of backup.

So how does a wind farm compare to a conventional fossil fuelled power station regarding generation capacity.

We will take, as an example, Aberthaw B coal-fired power station, which has an installed capacity of 1500 MW and assess how many large wind generators would be needed to compete with its output. Thus if we are to consider a large 2 MW wind generator with a load factor of 25 per cent and a load factor of 62 per cent, for Aberthaw B power station, then calculations will show we will need 1860 large wind generators (see Appendix 2) to compete with the capacity of Aberthaw B. Now this is truly mind boggling when considering the land area needed to accommodate 1860 large wind generators – you do not need much imagination to envisage the industrialisation and desecration of acres of local countryside to accommodate all these generators - it would be complete madness - and that is the situation for only one large fossil fuelled power station.

It is very important to remember also that these 1860 wind generators are totally dependent on the 'right kind of wind' whereas Aberthaw B power station will supply reliable and continuous electricity effectively at the throw of a switch - also that fossil-fuelled power stations are constrained by a number of controllable factors regarding their load factor such as human intervention (see Chapter One). Thus it can be argued the true figure for an equivalent number of wind generators will be greater than 1860 and it would not be exaggerating to assess the true number of equivalent wind generators at about 2000, or maybe more – but we will be generous and settle for 2000 – and it does not take a genius to decide which source of power generation, is without doubt, the superior option to be connected to for security of supply. To get a handle on the area of land required for 2000 large wind generators then a trip to a large onshore wind farm such as that of Llandinam, Powys, Wales, with a total of 103 wind generators would help – the generators here are not particularly large being only 31.7 metres (150 feet) tall. A visit to this site would mentally help to conjure up an image of the impact two thousand large 123 metre (400 feet) tall wind generators would have upon the landscape - a very enlightening but depressing experience.

A large coal power station will take up about 2 square kilometres of land, whereas 2000 large wind generators will require several hundred square kilometres, and as such, I do not think any right-minded person could countenance such a widespread desecration of countryside to accommodate so many wind generators - it really does beg the question, "Are wind generator supporters in complete denial?"

So let us see the equivalent number of wind generators that would be needed to compete with the following seven large coal fired power stations:

Station	Installed Capacity (MW)
Ferrybridge C	1955
Fiddler's Ferry	1961
Eggborough	1960
Cottam	2008
West Burton	1972
Rugeley	1006
Aberthaw B	1500
Total	12362

Again we will use the same load factor for a coal-fired power station at 62 per cent, and again a load factor of 25 per cent for a large 2 MW wind generator. Calculations will show (see Appendix 2) that 15,328 large wind generators would be needed. Thus to compete with seven large coal-fired power stations then over fifteen thousand 123 metre (400 feet) high 2 MW wind generators would be required - think about all those generators - remember 20 wind generators might extend over an area of 1 square kilometre – fifteen thousand large wind generators would require over seven hundred square kilometres of land - try and imagine this monstrous 'industrial impact' on the landscape – indeed the figures tell their own story.

To maintain power supplies without interruption, then it should now be obvious dear reader that wind farms will need backup by fossil fuelled and/or nuclear powered power stations in the UK. At the time of writing the British Wind Energy Association (BWEA) claimed that on average, 80 per cent of the public supported wind energy, with less than 10 per cent against it, and with the remainder undecided. They also stated that surveys conducted since the early 1990's across the country near existing wind farms have consistently found that most people are in favour of wind energy, with support increasing amongst those living closer to the wind farms.

Well, what can one say to these claims - I am truly amazed by this statement - I guess the population along the shoreline of the Bristol Channel will also be perplexed by the BWEA claim, especially when the proposed wind farm in that part of the UK during 2005 was halted due to wide protests. Therefore it is not surprising when I say I would challenge the BWEA claim any day as in my experience most people are against wind farms, particularly those who have had wind generators imposed upon them.

It cannot be stated enough times, especially in view of the outrageous claims expressed by the BWEA, that a wind generator to start producing any electrical energy it requires a wind speed in excess of 16 kilometres per hour (10 miles per hour). Additionally, it is important to have a sustained wind speed in excess (but not too excessive) of the quoted figure for any meaningful and continuous generation of electrical energy - to illustrate the limitations of these useless monstrosities I will offer the following personal experiences and observations which I feel highlights the utter nonsense of it all:

Tuesday 22nd November, 2005 proved a good day to illustrate the ineffectiveness of electrical wind power generation. Although a beautiful blue sky and the Sun shining above our village situated 500 feet above sea level in Pembrokeshire, the temperature remained at around minus 3 degrees Celsius. There was fog covering low lying ground and most of the eastern parts of the county, with little or no wind. Due to the fact that it looked so beautiful outside, my wife and I decided to wrap up warm and prune a number of hedges around the garden. By lunch time we had filled up our small trailer with the cuttings from the garden, so we decided to take them to our local refuge tip which caters for such waste. Our journey took us along the A478 which passes over the eastern edge of the Preseli mountain range and as such offered very good views across south Pembrokeshire. In a sense the view was quite surreal, with blue sky above but with a blanket of low fog hiding most of the low lying countryside - although one view really caught the eye above the blanket of fog – and that was the motionless and sentinel tips of the blades of a wind farm in the near distance - I can still clearly remember saying to my wife, "Look at that lot, what a waste of space, freezing fog, no wind and no power generation."

Wednesday 28th December 2005 was the coldest day in the United Kingdom for many years with temperatures dipping to minus 10 in many areas. Indeed, at 1800hrs (when there is a substantial demand for power), only 1 of the 70 Meteorological Offices was recording a wind speed of more than 16 kilometres per hour. Now think of the scenario if you are totally dependent on wind for the generation of electrical power. There would be no lighting, no television or radio (apart from battery operated), microwave ovens and electric stoves would prove useless, home computers would not function; most importantly there would be no heating. Remember, central heating systems, whether heated by solid fuel, oil or gas are totally dependent on an electrically operated system and circulation pump to function - thus you would be at home in the dark with no heating, no hot food or drink, no television or radio, with plenty of time on your hands to curse the 'beast' commonly and incorrectly called wind turbine.

If there is any doubt regarding the wind situation in the United Kingdom then I would suggest looking at Meteorological Office figures for wind speeds on the Internet. Some may argue that many Met Office stations are geographically irrelevant for judging wind power potential in Britain as the measuring equipment is much lower than the hub-height of a large wind generator – and although this may be true to a certain extent, the Meteorogical Office figures will give a good indication to the viability and effectiveness of wind power within the UK - the Renewable Energy Foundation (REF), a charity sponsoring research into the adoption of renewable technologies, announced in a press release on Friday 8th December 2006, that recent research using Ogfgem and Meteorological Office data to model output for every hour of every January from 1994 to 2006, indicated the volatility of a large installed capacity of wind power built across the UK – power swings of 70 per cent in 30 hours – the average January power variation over the period was 94 per cent of installed capacity and an uncontrolled variation decided by the weather. Thus the model indicated that even with the best efforts, large distribution of wind generators across the UK would have a low capacity credit and be difficult to handle.

Wind Generators are Dangerous

The developers of wind generators would have us believe their devices are now completely safe, even though they admit to past damage problems – but the Wind Farm Industry has no alternative to admitting to past damage as the historical record has provided the necessary evidence.

We are all aware of what storms can do to trees, houses and buildings, the power of the wind can indeed be very destructive, and it is instructive to quote a single example from the past:

A great storm raged across southern England in 1703 killing at least 8,000 people. The Eddystone Lighthouse disappeared without a trace along with Henry Winstanly, its designer. The average wind speed was said to be more than 100 miles per hour and, interestingly 400 windmills were blown down along with hundreds of churches.

Apart from the direct damage to structures caused by extremely high or gale force winds, there is also another consideration which may not at first be that obvious - we are all familiar with resonance and that it occurs when two objects naturally vibrate at the same frequency. Unfortunately, resonance can have its problems and that is why soldiers when marching across a bridge have to 'break step' - if the soldiers were marching in step and their steps coincided with natural frequency of the bridge, then the bridge could begin to vibrate violently, and in the worst case scenario would cause the bridge to collapse.

Wind can also set up unwanted vibrations in bridges and other structures. One of the most alarming examples of bridge damage due to resonance was the collapse of the Tacoma Narrows Bridge in the state of Washington, USA during 1940. Only just four months after the bridge was completed a mild gale caused a fluctuating force in resonance with the natural frequency of the bridge, steadily increasing the amplitude until the bridge collapsed.

Therefore it should come as no surprise that tall wind generator blades and their towers are also susceptible to this phenomenon. A very tall wind generator tower will swing backwards and forwards due to the wind. The frequency in which the tower moves backwards and forwards is known as the 'eigenfrequency' (the frequency with which a specific system may vibrate).

Obviously the amount of movement in a specific tower is dependent on factors such as the weight of the nacelle and its enclosed machinery, the propeller, the height of the tower, the construction material and thickness of the tower walls. Additionally, wind generator blades, by their very nature, tend to be flexible and therefore can vibrate with the wind. Now if the vibration of the wind generator propeller blade happens to coincide with that of the tower, then the oscillations can get out of control as in the case of the Tacoma Bridge, resulting in a devastating scenario.

Therefore it is vital that the eigenfrequency of both the propeller blade and tower is known to prevent such events happening. We can only hope that the designers and builders of wind generators are competent in their work and damage will not occur after installation.

The developers of wind generators will admit to past damage problems, but now claim the modern wind generator is completely safe.

So what are we to make of the following examples of reported damage to wind generators:

During the early part of April, 2005 a 40 metre blade from one of the state-of-art turbines situated at Crystal Rig, Lammermuir, Scotland shattered! Anders Falkjell, operations manager at Crystal Rig, confirmed the incident by saying, "It's true that one of them has broken and we are investigating that at the moment. It's not normal and I have not seen it before myself but I know that it has happened at other wind farms."

Quoting from the website, (www.windaction.org/posts/196-damaged-turbine-crystal-rig), David Bruce, of the pressure group Scottish Wind Assessment Project, said: "There were high winds so the turbines were 'feathered', or locked so they couldn't spin round. It was lucky nobody was walking below. This is only about the second incidence of this in the UK but it shows this is possible." For those who have computers which are connected to the Internet put 'crystal rig wind farm damage' into your search engine for pictures and a lot more detail.

The following is from (with acknowledgements) the Warmwell Internet site, but unfortunately is now closed, and I quote:

"January 1st, 2006 ~ Sunderland's spectacular display of 'safe' energy. Booker's Notebook. 'It was like a great flaming Catherine wheel," said a Sunderland resident just before Christmas, when flames engulfed one of the six 180-foot wind turbines recently installed at a cost of £2.3 million to provide just 5 per cent of the energy needs of the Nissan car plant. After a blaze visible for miles, the fibreglass blades crashed down into a field. It appears that the colossal confidence trick of 'safe', 'cheap', environmentally friendly, wind power is finally being exposed.

Plans to erect thousands more of these inefficient and expensive contraptions are arousing informed opposition throughout Britain. Meanwhile, new figures show that Denmark, which derives 20 per cent of its power from wind – the highest percentage in Europe – not only has Europe's highest electricity bills but has also fallen short of its Kyoto and EU targets for savings on carbon emissions by a staggering 25

If you think wind generators are safe, then you just might find your ideas going up in smoke!

per cent. The Advertising Standards Authority has just upheld a complaint against Renewable Energy Systems for exaggerating – by no less than 10 per cent – the emissions savings made by its turbines. Yet the formula used by the firm to arrive at these figures is the same used by the British Wind Energy Association and almost every other company seeking to build wind farms in Britain". End of quote.

During October 2013 there was a lethal wind generator accident in the Netherlands when a Vestas turbine caught fire and killed two workers –

from the European Platform Against Wind Farms (EPAW) website (www.epaw.org) and I quote with acknowledgements, "From our correspondent from The Netherlands: two wind turbine mechanics, respectively 19 and 21 years old, died because of the fire. One fell to his death and was found on the ground underneath the turbine, the other died from his burns and was found inside the charred remains of the turbine.

A surprisingly large number of wind turbines are involved in accidents around the world. Most of them are blades falling off, turbines collapsing, or nacelles burning down to a skeleton (400 - 800 litres of burning oil are not easy to extinguish, especially as firemen rarely have ladders long enough for these >100-meter long contraptions). Some human deaths have been reported.

A record is being kept at Caithness Windfarm Information Forum.

Some bush and forest fires have been caused by wind turbines, but such news have never surfaced, or have rapidly disappeared from the radar screen. Here is one: Cal Fire: Wind Turbine Generator Caused Wild land Fire that Charred 367 Acres. Indeed, it's not politically correct to report such things." End of quote.

I could fill this book up with numerous examples of such reports relating to wind generators, and for readers who wish to know more should visit the website run by Angela Kelly at (www.countryguardian.net).

There are also many other sites on the Internet that will feed your appetite should you so desire - just put something like, 'damage to wind generators' into your search engine - and be enlightened.

Lightning

It is perhaps difficult to believe but there are approximately nearly 2000 thunderstorms occurring across the globe at any one instant. Indeed, the thunderstorm is the most common type of storm - such countries as Nigeria experience a thunderstorm nearly every two days. Lightning, apart from killing people, can start forest fires, ruin crops, set property on fire - with sensitive electrical equipment such as computers being vulnerable from power surges when lightning strikes.

I wonder how many folk have ever considered why forked lightning follows such a seemingly tortuous path from cloud to ground when the obvious shortest route is a straight vertical path to the ground - in simple terms, it is because the electrical discharge is attempting to follow the path of least electrical resistance between the air and ground (earth).

This is one reason why lightning protection can sometimes fail – take a church steeple as an example, the lightning rod should be at the highest point, whilst the copper conductor should run as straight as possible from the top lightning rod to the ground earth system – lightning does not like bends - any unacceptable bends, then the lightning discharge will find another but easier path to earth.

Wind generators have lightning conductors embedded in their blades, as well as down the tower. Now should lightning strike a blade, this begs the question of how the lightning discharge passes easily (path of least resistance) to the ground – remember lightning does not like tortuous routes, such as from blade to tower and hence to ground.

Most of us are aware that to be in an open area such as on a golf course, or on top of a hill, during an electrical storm can be very perilous, and no one in their right mind wishes to be struck by lightning - the electrical potential of a lightning strike can be in the order of 100 million volts and the current can be up to 200,000 amps - although an average value is nearer 25,000 volts - imagine the potential damage that can occur to a hill top wind generator that is not adequately protected.

Cloud to ground strokes may cause currents of thousands of amps to flow along wires or cables for microseconds only, but still result in many thousands of volts being applied to plant causing damage. It should be noted though that it does not need a direct strike to cause damage, as a nearby ground strike can raise the ground potential to several thousands of volts. Thus any cables laid in the ground, be they electrical or telephone can be subject to high electrical stresses (ground to inner cable conductors) and if not adequately protected can indeed fail.

My experience as a telecommunications engineer has taught me that a significant lightning strike does not take any prisoners - I have witnessed telephone lightning protection equipment literally blown off walls and telephones ending up in almost a molten mass.

According to the National Grid earth wires 'trap' about 90 per cent of lightning strikes to their overhead lines – if you view a high voltage pylon route you will see a single earth wire strung between pylons and attached to the very top of the towers - so it follows that only 10 per cent of lightning strikes to their lines actually hit the phase wires. National Grid say the effectiveness is a function of the design of the tower and the arrangement of the phases and earth wire. The design of pylons seen across the country, reflect the fact that the UK is not a particularly severe region with respect to lightning, when compared to some parts of the USA or Brazil for example. More severe regions would necessitate a different design of tower, often including a double earth wire. National Grid say that to put things in perspective, their specifications (NGTS 1) ask for a line design that results in a maximum of one event (phase hit) per 100 kilometre per annum.

To enable an overvoltage to be grounded during a phase strike on UK Grid lines, arcing horns with a specified gap are connected across the line insulators - the air gap is the 'weakest link' of the insulator set, so any overvoltage should flash across the gap, and not through the insulator string; exceptions to this are typically when the dishes are contaminated by pollutants, or defective, resulting in an alternative 'weakest link' for the flashover arc.

The reader may not be aware that cows have been found dead in fields in close proximity to high voltage electricity pylons. This can be the result of either lightning hitting the actual power lines of grid routes causing an 'overvoltage', which then flashes across the arcing horns of the dish insulators, allowing the high voltage to flow to ground (earth), or due to direct strikes to the pylon itself which are naturally grounded. Depending on the resistivity of the local area, ground potential can be of a very high order such that the potential gradient between the front and rear legs of the cow becomes fatal, resulting in the death of the animal. It does not require a cloud to ground to cause damage to electrical plant as cloud-to-cloud discharges can produce large electromagnetic pulses that can induce high voltages into electrical equipment. Considering the potential damage lightning can possibly do to a single wind generator on top of a hill it is rather sobering to imagine the amount of damage that could happen within the area of a wind farm and its numerous generators, especially during a violent thunder storm.

Thus with regard to wind farms in the UK, visitors, hikers and walkers should be fully aware that the messages given in warning notices, which are placed at entrances to wind farms, should be rigorously adhered to. The following warning notice is at the Mynydd Clogau wind farm near Newtown, Powys and reads as follows:

VISITORS WEATHER WARNING

ICING

During wet and freezing conditions there is a risk that ice can form on the blades on the wind turbine.

Please DO NOT ENTER the wind farm in these conditions.

LIGHTNING

Please DO NOT ENTER the wind farm during lightning conditions.

If a lightning storm starts during your visit please leave the site IMMEDIATELY.

If you require assistance please call 01686 629803

THANK YOU FOR YOUR ASSISTANCE

To ignore the warning could possibly be dicing with death - during wet and freezing conditions, or that of an electrical storm the last place anyone should be walking through is a wind farm; not just in fear of the possibility of debris falling, but also to the potentially dangerous nature of the ground upon which they would be walking.

When lightning strikes a wind generator it will hopefully travel down the lightning rod/conductor and discharge into the ground; now depending on the magnitude of a lightning strike, ground resistivity, and the effectiveness of the generator earth system, a dangerous, if not fatal, scenario can arise, known in engineering terms as a *rise-of-earth-potential* situation.

This means that during the duration of a lightning strike, a voltage (potential) gradient will appear in the ground, diminishing in magnitude whilst spreading outward from the wind generator earth electrode system. Anyone foolish enough to be walking on this ground at the time of a strike will have a voltage (potential) difference appearing between their feet, which, as mentioned above can prove fatal - the situation is worse for animals whose legs are spaced further apart such as dogs or sheep - indeed, horses and cattle are at even greater risk due to the larger spread of their legs and hence larger potential difference.

It is worth reiterating that cows have been found dead near electricity pylons due to *rise-of-earth-potential* situations - now wind generators are no different to electricity pylons in this respect - it can be argued they pose even a greater threat, due to their very nature, being built on high exposed ground, which usually has high resistivity - the greater the resistivity, the greater the problem - not forgetting that the higher the ground and the taller the wind generator, the greater the exposure and risk to lightning strikes.

According to the National Lightning Safety Institute, Louisville, USA and I quote: "The USA Wind Farm Industry (WFI) largely is centred in low-lightning areas of the state of California. While some evidence of lightning incidents is reported here, the problem is not regarded as serious by most participants. The USA WFI now is moving eastward, into higher areas of lightning activity. The European WFI has had many years of experience with lightning problems. One 1995 German study estimated that 80 per cent of wind generator insurance claims paid for damage compensation were caused by lightning strike. Neither the European, or USA WFI has adopted site criteria, design fundamentals, nor certification techniques aimed at lightning safety. Such guidelines are necessary if lightning hazard reduction at wind farms is to be an achievable goal." End of quote.

Various case studies on the American National lightning Safety Institutes Internet site at www.lightningsafety.com make for interesting reading, such as: Eighty-five per cent of the downtime experienced by a second south western USA commercial wind farm was lightning related during the startup period and into its first full year of operation. Direct costs were some $55,000, with total lightning-related costs totalling more than $250,000.

Closer to home a 1996 European retrospective study was conducted of some 11,605 wind generators in Denmark and Germany, and accurate operational records were available for analysis. General findings indicated:

a) Lightning faults caused more loss in wind turbine availability and production than the average fault.

b) Ranking in descending susceptibility to lightning damage were turbine control systems, electrical systems, blades, and generators.

c) The number of failures due to lightning increases with tower height.

d) Wood epoxy blades have significantly less damage rates than GRP/glass epoxy blades.

Regarding the exposure of wind farms in the UK to lightning, the wind generator manufacturers/installers will claim, that if they get their 'lightning protection' correct they can meaningfully limit most potential damage. But I would suggest reader's put something like, "Lightning damage to wind generators" into their computer search engine and then decide for themselves.

Earth Leakage Voltage

Another potentially dangerous factor to consider when visiting a wind farm is the appearance of a stray voltage on the surface of a structure that houses electrical equipment, and of course, across the surface of the ground at any point in the vicinity of electrical equipment which becomes faulty – this is usually caused by the breakdown of the insulation of electrical windings such as in the case of a transformer or electrical generator, not forgetting the breakdown of insulation within a power cable – this stray voltage is known as an *earth leakage voltage*.

Similar to the situation of the *rise-of-earth-potential* due to a lightning strike, the ground potential can be of a lethal nature, and unlike a momentary lightning strike the earth leakage voltage can last for quite a while, making it very dangerous indeed i.e. if the integrity of the protective earth electrode system is violated then the voltage may persist for some considerable time due the failure of a circuit breaker to open or a fuse to blow!

The reader should be aware that all electrical equipment are bonded by a conductor to an earth electrode buried in the ground, which usually takes the form of a large metal plate or several copper rods connected together and buried in the ground – the size and depth of a plate and/or rods is determined by local ground resistivity – the higher the resistivity the larger the plate and/or number of rods and the deeper they are all buried.

In a wind generator, the generator case, for example, is normally bonded to a copper tape earth conductor leading down the tower to a suitable earth electrode buried in the ground. It should be remembered that wind generators by their very nature are placed as high as possible such as on top of hills – but unfortunately, this is the very place where ground resistivity can be high, making it difficult in providing a quality and fully effective earth electrode system.

Electrical engineers are well trained to avoid any hazards from stray earth leakage voltages, being well aware to keep hands, unless adequately protected, well away from live electrical equipment. Indeed, in power stations, staff will stand on glass or ceramic floors which, because of their insulation properties, offer a high degree of protection. Water, in this respect, is the enemy of electrical engineers and you are virtually taking your life in your hands by touching any high voltage electrical equipment whilst standing on wet ground – those warning notices around high voltage pylons and transformers are not put there for fun.

The lesson from all this - unless you are well trained and know what you are doing - is to keep well away from wind generators and certainly do not attempt to hug one, or any electrical equipment that is in close proximity.

TV Signal Interference

By the very nature of their structure wind generators can and do interfere with terrestrial transmitted television signals – but of course it depends where the receiving aerial is in relation to the TV transmitting mast and the wind generators – out of the 'shadow' of the wind generators there should not be any problems. Where such issues arise though, there are a number of options available to overcome the problem.

Three 1.3 MW wind generators at Blaen Bowi, Carmarthenshire in West Wales provide a good example where local residents found that after the wind generators had been installed, they experienced interference with their television reception. It should be noted the main transmitter for Blaen Bowi is named Presely, and is located at Pentre Gala having a horizontal polarised signal - the existing Llandyfriog relay station has a vertical polarised signal.

The following is an extract of a case study from the Sustainable Development Commission (SDC) website at (www.sd-commission.org.uk) and is reproduced with their kind consent:

Windjen Power Ltd commissioned Crown Castle UK, the primary broadcast transmission company in the UK, to carry out a survey to establish what effect the turbines would have on TV reception. Whilst not 100 per cent conclusive, the survey identified that a repeater would be required on the mast, at the Llandyfriog relay station.

The report also stated that the full effect on TV reception would not become apparent until the wind farm was operational.

A local electrical/aerial engineer was employed to remedy interference issues in July 2002. Some 26 households from surrounding villages did report problems with their TVs. All reports of TV interference up to an 8-10 mile radius were investigated and dealt with as they were received. Rectifying the interference on analogue signals for television took between nine to 12 months. An added bonus is that some unaffected families now receive Channel 5. The planning authority agreements stated a limit of £5000 that the developer was required to spend on the problem. Windjen Power Ltd, based in Colwyn Bay has spent £33,000 in resolving TV problems.

D. Jones, Managing Director, Windjen Power Ltd, stated that, "Problems of this nature can be quickly resolved given the understanding we have gained. Measures can also be put in place to minimise the TV reception interference after wind farm commissioning."

The case study showed how simple solutions can often be found to problems such as TV interference, and that good consultation with local residents is essential. Repeater transmitters are just one of a number of options available to developers. In some cases, the installation of satellite TV at affected households is an alternative option.

Electrical Generation in Wales

There is absolutely no technological, engineering or environmental sense in the provision of onshore, sporadic, electrical wind generation in Wales, or indeed, for offshore generation - this is because of the following inescapable facts:

- ❖ Wales, at the time of writing, generated a total of 33.5 terawatt hours (TWh) of electrical energy per annum - of this total amount the principality consumed only 17.6 TWh - the surplus (15.9 TWh) being fed into the National Grid for use elsewhere - so just under 50 per cent of Welsh generation is fed to England.

- ❖ Prior to 2008 the amount of electricity generated in Wales remained relatively stable with around 35,000 GWh generated per annum (wales.gov.uk). However in recent years the amount of electricity generated in Wales has been falling, with 27,300 GWh generated in 2011, see Appendix 5.

- ❖ The relatively new 2000 MW gas-fired power station in Pembrokeshire is the largest and most efficient in the UK, not forgetting other power stations such as Aberthaw, Connah's Quay, Usk Mouth, and the Rheidol and Ffestiniog Hydro-power stations, all contribute (because of the Grid) to exporting Welsh power to England. This is not to overlook Dinorwig pumped-storage power station housed in 'Electric Mountain' at Llanberis. This power station was built to satisfy high power demand from predominantly English homes – for example, when electric kettles are put on after an eventful programme on the television. So how much more must be asked of the Welsh people - do they also have to sacrifice the beautiful Welsh countryside and tourist industry for limited and unpredictable wind.

- ❖ Wind generators by their very nature are totally dependent on the vagaries of the wind and as such have load factors of 29 per cent and less - remembering that with very little or too much wind there is no generation - the variable nature of their output can cause difficulties with the Grid - and of course for security of supply require backup by conventional power stations.

- ❖ It is nonsense claiming that a wind generator when running produces no emissions, when indirectly it does - every wind farm necessitates the backup from a conventional power station.

- ❖ Informed people are fully aware that a 'hundred thousand wind generators' in and around Wales will have nil effect on climate change - to claim otherwise simply demonstrates ignorance of climatology, wind technology, and the global situation.

- ❖ Wales relies heavily on tourism and the mindless and useless industrialisation by the construction of these ineffective monsters is disgraceful and a betrayal to the people of Wales and future generations.

Problems in Local Power Networks

With the growing number of significantly sized wind generators being built on numerous farmlands and connected directly to the local electricity network, then overloading will become a *real* problem, resulting in possible brownouts, and in the worst case, power blackouts – the problem will also be exacerbated by the significant number of private dwelling roof mounted solar PV arrays, and the steadily increasing number of solar parks – unless, of course, the power distribution companies are effective in upgrading their networks to meet the situation – then, of course, there is the question who will pay for such works?

The reason the integrity of local power networks will be threatened with overload, is simply because the Grid and local power networks have been designed (with known capacity) to supply electricity from central (monitored) sources (power stations) to the customer, and not the other way around - from the customer back into the network. As more and more wind generators (and solar panels) are connected to the local networks (exceeding planned capacity), and if the local power distribution companies are slow in upgrading the necessary external line plant, then problems will inevitably arise.

From Chapter One you may recall that a typical example of a 132 kV Grid to local network system may consist of a power station generating 25 kV, being transformed up to 132 kV for transmission over the Grid, to a transformer station dropping the 132 kV down to 33 kV and 11 kV for distribution to transformer dropping the 11 kV down to 415/240 volts for distribution to the consumer. Please note that the accepted allowed voltage at the household is between 216 Volts to 253 Volts. The same basic principle applies for a 400 kV Grid system.

Large wind farms that are directly connected to the Grid are managed and monitored in much the same way as conventional power stations – such that when National Grid perceives there is sufficient power in the Grid and that too much energy will be generated by the directly connected wind farms, then some or all of these wind farms are asked to shut down to avoid overloading the National Grid - under the 'constraint' payments process owners of wind farms can apply for compensation for each megawatt hour of energy their wind farms would have produced. It should be noted though that if a number of large wind farms are planned then enhancement of the Grid may be required to accommodate them.

Electricity demand will rise sharply after major events or following the climax of a popular television programme when a large number of viewers collectively return to everyday business, including power-consuming habits such as switching on lights, computers - or the kettle. Therefore National Grid needs to forecast demand and supply as precisely as it can to prevent blackouts, which can result from sudden surges placing a big strain on the electricity network.

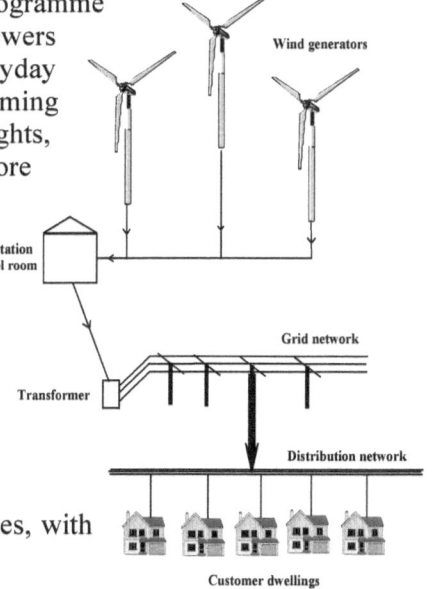

The UK has, in the past, seen huge jumps in electricity consumption during football games, with highest ever spike of 2,800 megawatts - equivalent to 1.1 million kettles - recorded after England lost the 1990 World Cup semi-final penalty shootout against West Germany. Such events can generally impose an extra demand of 200-400 megawatts (MW) on the Grid – this short term demand is usually met by pumped storage schemes such as Dinorwig in North Wales, which can be quickly brought online.

Simplified diagram of a wind farm showing one example of its power distribution. Electrical energy generated from the wind generators is fed, via the control room, and the step-up transformer, to the local Grid. Once in the Grid the energy can be fed, via step-down transformers, to the distribution network. The customer then taking power from his distribution network. Depending on the situation the electrical energy from the wind farm could be fed direct to a local distribution network, or indeed, directly to a large energy user.

But with local networks the situation is not that simple - the unpredictable input from the ever increasing number of relatively large wind generators we now see 'popping up' around the countryside, in addition to household solar arrays and solar parks, means that the existing local network may not be robust enough for this additional load, and thus require upgrading. Envisage a winter scenario when many householders, after a popular programme on the television, head to the kitchen to put the kettle on - this will result in a power surge in the local network – then at the same time a strong wind suddenly starts to blow so that local wind generators will be inputting their maximum capacity into the network – unless the system has been suitably upgraded to accommodate this overvoltage it will stressed to a point where problems will arise.

An example of what can be happen due to overvoltage conditions occurred on Sunday 8th March, 2015, at our property in Ceredigion. In fact I was working on this book when I became aware of something amiss at about 1500hrs noticing my laptop (monitoring the house solar panels), had flagged up a Grid problem. This prompted an inspection of the solar PV system inverter situated in the garage - which I found had shut down, showing a red LED light and a display message, 'Grid Fail – Trying to Connect'. Later when I looked at the graph display on my laptop it indicated the problem had lasted for two hours, from 1400hrs until 1600hrs – this obviously meant that our solar panels, which had automatically shut down, did not provide any electricity for two hours. Checking the grid monitoring and device settings for the inverter later on my laptop, I noted the inverter had been set to cut out at 264 volt – this meant that at the actual time of the inverter cutting out the mains voltage had to have reached at least 264 Volts.

Luckily I had had the foresight to have a voltage regulator installed at the same time as the solar panels and this device prevented other household sensitive electrical equipment, such as the TV and computer, being exposed to electrical stress and possible failure – noting that electrical stress can result in component failure at a later date, other than at the time of the event - the advantages of a voltage regulator are further explained in Chapter Seven. Electrical installers of wind generators and solar panels all have to forward the necessary information to the power distribution companies for authorised connection to the power network. So hopefully the power distribution companies receiving this data will, at an early date, carry out the necessary assessment and any necessary works to overcome the possible problems as foreseen by the author – and not wait until problems arise before acting!

Wind Generation Subsidy

It may come as a surprise to some folk that wind generators receive significant government subsidies – readers should also be fully aware, as mentioned above, that when the National Grid perceives that too much energy will be generated, the directly connected wind farms are asked to shut down to avert overloading of the National Grid, and under the 'constraint' payments process owners of wind farms can apply for compensation for each megawatt hour of energy their wind farms would have produced. It should be noted that at the time of writing the owners of 10 wind farms have been paid more than £3 million each to shut down when the Grid has not required electricity produced by wind energy with the highest payment of £11.1 million being paid to Scottish Power. Surely all of this has to be the engineering and economics of the madhouse encouraged by pusillanimous political minds that do not have a clue, with the unfortunate consumer footing ever increasing energy bills coupled with an ever increasing prospect of a power failure.

Hopefully having read so far the reader will now be fully aware of the lunacy of large scale wind generation in the UK and may be feeling somewhat aggrieved to having the wool pulled over their eyes and having their pocket picked– and just to put some salt into the wound, as they say, the reader should also be aware that none other than Paul Golby, the chief executive of E.ON UK (formerly Powergen) has said, and I quote, "Without the renewable obligations certificates nobody would be building wind farms" end of quote. Unless there is a drastic change of strategy from government, then it would seem prudent to invest in that petrol or diesel stand-by generator in preparation for when the lights start to go out.

Summary

We have seen that fundamentally, the incorrectly named 'wind turbine' consists of no more than a large propeller with its shaft connected directly to, or via a gearbox and clutch, to an alternating current generator - it is simply a WIND DRIVEN GENERATOR, no more, no less! These generators being totally at the mercy of the wind, with load factors of 29 per cent, or less. Wind generators also require backup to ensure continuity of supply. If the backup consists of fossil-fuelled power stations then the claim to the saving of greenhouse gasses is obviously a much exaggerated, or indeed, a false one. Backup by nuclear power stations is also not a 'green' option due the problem of nuclear waste - not to mention nuclear mishaps such as Three Mile Island, Chernobyl, et cetera, see Chapter Five headed, Energy Alternatives, later in this book.

The evidence is also showing that wind generators are not as safe as the Wind Industry would have people believe, with plenty of images on the Internet of these monsters catching on fire, and indeed of collapsing.

Then there are the problems during electrical storms - they are not placed, regarding potential lightning damage, in the safest of positions on tops of hills and in wide open spaces. The positioning of a single large wind generator is cause for concern, but with the on-going proliferation of wind farms and their subsequent impact on the landscape the incidence of damage obviously increases - areas such as Wales, Scotland and many parts of England will pay a heavy price in the desecration of the local countryside and the almost certain loss in tourism - who will want to visit an industrialised landscape - a once peaceful and natural vista - now cluttered with these ugly, whirring monsters. I have to fully agree with the conclusion of Angela Kelly's article in the magazine Green Places when it states and I quote:

"Good planning is about balance. The irreparable ecological damage, loss of amenity and distressing divisions within communities caused by commercial wind turbines far outweigh any benefit of their insignificant and unreliable contribution to our energy needs. The tiny, intermittent output of electricity and the negligible CO_2 savings cannot possibly justify the huge sacrifice of that most finite resource – our unspoilt and unrenewable countryside. It is our duty to protect our rural heritage for present and future generations from such gross and unnecessary industrialisation. The alternative will be a national disaster – no less." End of quote.

Regardless of what the Wind Industry claim, people living near these monstrosities do suffer from noise and strobe effects, again this can be verified by researching local papers and logging onto the Internet and visiting Country Guardian and other such sites.

The reader should be aware that the Climate Change Act (2008) requires the UK to cut its CO_2 emissions by 80 per cent from 1990 levels by 2050. This will come at a horrendous cost resulting in such electricity driven industries as Aluminium production shifting to China, where of course Aluminium will then be shipped back to the UK exacerbating global CO_2 emissions – what a brilliant strategy – remembering also that most electricity in China is generated by coal.

Thus it would seem as if the lunatics have taken over the asylum - the evidence is there for you to see, if you are willing to take the time to look for it - then lobby your local councillor, MP and AM about this total madness.

You owe it to yourself, but most importantly to your sons and daughters – the next generation.

To conclude it is wise to note the motto of the Royal Society (founded in 1660):

Nullius in verba

Take nobody's word for it, check it out for yourself.

"Let all men know how empty and worthless is the power of kings. For there is none worthy of the name but God, whom heaven, earth and sea obey."

King Cnut (Canute) the Great (995 – 1035)

CHAPTER FOUR

Ocean Tides

I wonder how many readers have really questioned why there are tides on our planet – indeed, not many people appreciate there are also land tides, or know that there are atmospheric tides - but apart from long lost tribes deep in remote jungle areas, we are all aware of the ocean tides, with some folk being much more acutely aware to the nature of the tides having been caught out by a rising tide. Yet it is enigmatic that very few can fully explain the mechanism behind tidal  movements saying such things as they are due to the influence of the Moon without really understanding what these words mean, and if pushed to explain further, will say it is due to the gravity of the Moon pulling on the oceans, and again not really understanding what their words mean…

So we will begin by stating straight away why we have ocean tides on the Earth:

The ocean tides are the result of the Earth falling toward the Moon and to a lesser extent of the Earth and Moon system falling toward the Sun.

For those unfamiliar with the forces behind the tidal movements these will indeed seem strange words and possibly protest that the Earth does not fall toward the Moon, neither does the Earth and Moon fall toward the Sun, saying, "That's nonsense!"

Therefore before proceeding further and to fully understand the forces behind the tides it is necessary to be re-acquainted with Isaac Newton (1642 – 1727) and his Universal Law of Gravitation, also to renew our understanding of what is meant by inertia, mass, momentum, force, weight, acceleration, acceleration due to gravity, centripetal force, barycentre and free fall.

Newton's Universal Law of Gravitation

Newton's law of Universal Gravitation states that every mass in the Universe attracts every other mass with a force that for two masses is directly proportional to the product of their masses and inversely proportional to the square of the distance separating them. When a constant known as the universal gravitational constant (G) is introduced we can describe Newton's law of gravity in a simple equation:

$$F = \frac{G m_1 m_2}{d^2}$$

Where F = force of gravity between two objects, m_1 and m_2 (newtons).
m_1 = mass (kilograms).
m_2 = mass (kilograms).
d = distance between the centre of the two masses (metres).
G = Universal gravitational constant. (6.67 x 10-11 $N\text{-}m^2 / kg^2$).

The greater the masses m_1 and m_2, the greater the force of attraction between them. The greater the distance (d) of separation, the weaker the force of attraction.

It should be noted that the force weakens as an inverse-square law. The inverse-square law is a law relating the intensity of an effect to the inverse square of the distance from the cause. Gravity follows an inverse-square law, as do the effects of electric, magnetic, light, sound and radiation phenomena. Since the force of gravity decreases as the square of the distance, a planet twice as far from the Sun is pulled toward the Sun with a quarter the force; three times as far, a ninth and four times as far, a sixteenth and so on. Thus the pull of gravity diminishes very quickly with distance and if plotted graphically would follow an exponential curve.

Inertia

Newton's first law (Principle of Inertia) states that all bodies preserve their state of rest or their state of uniform motion in a straight line, except in so far as it is made to change that state by external forces. Now every material object possesses inertia: how much depends on the amount of matter in the substance of the object. Inertia (inertial force) is measured in newtons within the metric system, (it should be noted that in the imperial system the unit of force was measured in pounds). We frequently meet examples of the inertia of matter in daily life. If you are standing up in a train, which suddenly starts to move forward you fall backwards because some force urged your feet forward.

But the upper part of you, not being acted upon by a force, tended to remain still. If you are standing up in a moving train, which suddenly stops you fall forward for a similar reason. Jumping off a moving bus in the same direction as the bus is moving can be deemed dangerous, unless you deliberately throw your body backward, in order to land in a forward slanting position when you reach the ground. The inertia of a flywheel keeps an engine running smoothly. Some bodies have more inertia than others; in other words, it is more difficult to start them from rest or stop them in motion. It is easier to push a small car rather than a lorry along a level road; the reason for this is because there is a greater quantity of matter in the lorry than in the small car. In other words the mass of the lorry is greater. This is not to be confused with the weight of the lorry as we are not trying to lift it. Therefore the greater the mass of a body the greater will be its inertia. Indeed, some scientists regard inertia as just another form of mass.

Mass

Mass is the amount of material (matter) in an object. Matter is anything which occupies space and is classified into solids, liquids and gasses. The greater the mass of an object the greater is its inertia or resistance to change. Therefore mass is the tendency of an object to resist being moved, or if it is moving, to resist a change in speed or direction.

Mass is measured in kilograms: The standard unit of mass, the kilogram, is a block of platinum preserved at the International Bureau of Weights and Measures in France. The kilogram equals 1000 grams. A gram is the mass of 1 cubic centimetre (cc) of pure water at a temperature of 40 degrees Celsius. (It should be noted that the standard pound is defined in terms of the standard kilogram; the mass of an object that weighs 1 pound is equal to 0.4536 kilograms).

Momentum

The momentum of an object is defined as the product of its velocity and mass and its unit is the kilogram metre-per-second, (kgm/s). A heavy lorry is much harder to stop than a small car, both moving at the same speed - thus the heavy lorry has more momentum than the small car.

Force

A force is a push or pull upon a body resulting from the bodies' interaction with another body. Whenever there is an interaction between two bodies, there is a force upon each of the bodies. When the interaction ceases, the two bodies no longer experience the force. Forces only exist as a result of an interaction.

Forces can exert a push or pull at-a-distance despite their physical separation. Examples include gravitational forces - the Sun and planets exert a gravitational pull on each other despite their large spatial separation – when you jump in the air and you are no longer in contact with the ground, there is a gravitational pull between you and the Earth. Electric forces are at-a-distance forces - the protons in the nucleus of an atom and the electrons outside the nucleus experience an electrical pull towards each other despite their small spatial separation. Magnetic forces are at-a-distance forces - two magnets can exert a magnetic pull on each other even when separated by a distance of a few centimetres.

Suppose a body to possess a certain momentum, then for the momentum to change or tend to change, something must act upon the body and that something is termed force. It should be clearly understood that by thus defining force we do not get to know anything more about it. All we know are the effects produced by a something we call force.

A change of momentum is produced by force; the rate at which the momentum changes may therefore be used as a measure of force. Therefore the unit of force can be defined in several ways: A unit of force acting for the unit of time is able to produce a unit of velocity in a unit of mass. Or, a unit of force produces a unit of acceleration in a unit of mass. But since the product of mass and its velocity is spoken of as the momentum of the body, we can measure force by the momentum it generates, that is, the unit force giving rise to the unit of momentum in the unit of time.

Equal forces are therefore those that produce equal momenta in equal times.

The momentum generated by a force of two units is twice as great as that produced by one unit. Further, a force of one unit acting for two seconds will produce twice the momentum that it would do if it only acted for one second. This is why it is necessary in defining the unit of force to introduce the words 'acting for the unit of time'.

The momentum of any particular body is determined by the bodies mass and velocity. As the mass of the body may be regarded as constant any change of momentum can only be produced by changing the velocity. But rate of change of velocity is acceleration. Hence when the acceleration of a body is altered, the momentum is altered. An alteration of momentum signifies, as has been explained, that a force is acting upon the body. If the acceleration is uniform, then a uniform force must be acting upon the body.

The relation between force, mass and acceleration may be expressed algebraically as follows:

$F = ma$

Where F = force in newtons.
        m = mass in kilograms.
        a = acceleration in metres per second per second.

Weight

The downward force that results from the gravitational attraction of the Earth to bodies on its surface is called weight. It is directly proportional to the mass of the body and the mass of the Earth, decreasing inversely proportional to the square of their distance apart. To put it more succinctly weight is the 'force acting upon an object due to gravity'. Thus an astronaut will experience different weights on celestial bodies of different mass. For example, on the Moon, which has less mass than the Earth, the gravitational force is only 1/6 as strong as on the Earth. Thus a person weighing 80 kilograms on Earth will only weigh approximately 13 kilograms on the Moon.

Weight is measured in newtons, where 1 newton is the force required to accelerate 1 kilogram at 1 metre per second per second. The unit is named after Sir Isaac Newton (1642 – 1727). It should be noted that in 'everyday usage' weight really refers to the mass of a person or object. That is why people, when asked their weight, reply in kilograms, (or stones, pounds and ounces).

The difference between mass and weight must be clearly understood. Weight will vary dependent upon the mass of the bodies. On more massive planets astronauts weigh more than on less massive celestial bodies. Weight is a force. But the mass of the astronaut (unless he has been on a diet) will remain the same. Mass is the quantity of matter in a material body. The astronaut offers the same resistance to speeding up or slowing down regardless of whether the Earth, Moon, or anything at all is attracting him.

For example: In a spaceship located at a point between the Earth and the Moon, where gravitational forces cancel each other, the astronaut still has mass - if he were to stand on a scale, he would not weigh anything, but his resistance to a change in motion is the same as on Earth. As a further example it must be recognised that it would take the same push (effort) to start moving a vehicle on a level surface on the Moon as it would on the Earth, but it would require much more effort to lift the vehicle on Earth that it would on the Moon. The extra effort is required because of lifting the vehicle against the force of gravity. Mass and weight are very different from each other.

We have stated that the relation between force, mass and acceleration may be expressed algebraically as:

$F = ma$

Therefore in determining the weight of a body we may use the following formula:

$W = mg$

Where W = weight in newtons.
        m = mass of body in kilograms.
        g = acceleration due to the Earth's gravity ($9.8 m/s^2$).

As an example we can use the above formula to calculate the weight of a 1 kilogram mass when it is on the surface of the Earth:

W = 1 x 9.8
$$= 9.80 \text{ newtons}$$

Thus a mass of 1 kilogram will have a weight (force) of 9.8 newtons (2.2 pounds) acting upon it at the surface of the Earth due to gravity.

Acceleration

Acceleration is the rate at which an object's velocity changes with time; the change in velocity may be in magnitude (speed), direction or both.

Acceleration Due to Gravity (g)

Acceleration due to Gravity (g) is the acceleration of a free falling object under the effect of the Earth's gravitational field. Its value near the Earth's surface is about 9.8 metres per second each second. Thus we can derive an expression that shows how the acceleration due to gravity varies with distance from the centre of the Earth at points exterior to the Earth's surface:

$$F = \frac{Gm_e m_o}{d^2}$$

We also know that force equals mass multiplied by acceleration.

$$F = m_o g$$

Now let,
m_e = mass of the Earth (6×10^{24} kilograms).
m_o = mass of object. (kilograms).
d = distance between object and centre of the Earth (metres).
r = radius of the Earth (6.38×10^6 kilometres).
G = Universal gravitational constant. (6.67×10^{-11} N-m²/kg²).
g = acceleration due to gravity on Earth. (9.8 m/s²).

Thus we have,
$$m_o g = \frac{Gm_e m_o}{d^2}$$
$$g = \frac{Gm_e}{d^2}$$

Therefore an object 6000 kilometres above the Earth's surface would have an acceleration (g) of:

$$g = 6.67 \times 10^{-11} \times 6.0 \times 10^{24} / (6.38 \times 10^6 \times 6.38 \times 10^6)^2$$
$$= 4.02 \times 10^{14} / (1.238 \times 10^7)^2$$
$$= 2.60 \text{ m/s}^2$$

Centripetal Force

Centripetal means 'centre seeking' or 'toward the centre' and centripetal force is any force that causes an object to follow a circular path. Examples are electrical and gravitational forces: The orbiting electrons in atoms experience an electrical force toward the central nuclei. It is centripetal force, i.e. the gravitational force of the Sun that keeps the Earth in orbit around the Sun. If this force should suddenly fail then the Earth would shoot off into space at a tangent to its orbit at approximately 107,320 kilometres per hour, (66,700 mph). It should be clearly understood that centripetal force is not a separate force as such, but a generic term for all centre seeking forces.

Barycentre

Most celestial bodies are symmetrical in that they are spheroid and the centre of mass is at the geometric centre of the body.

Thus a freely rotating body will always rotate about an axis passing through its centre of mass. Now two or more bodies, such as a multiple star system or a star and its accompanying planets, will have a common centre of mass. Binary stars orbit about their common centre of mass, which in astronomical terms is known as the Barycentre.

If two stars of the same mass are gravitationally linked then they will orbit equally about their common centre of mass. If one of the stars is greater than the other, then the common centre of mass is shifted towards the more massive star. If one of the stars is sufficiently massive enough then the common centre of mass will reside within the larger star; this is exactly the situation with the Earth and the Moon.

The mass of the Earth at 6×10^{24} kilograms is far greater than the Moon at 0.0735×10^{24} kilograms; indeed it is 81 times as more massive. So the centre of mass for the Earth, Moon system lies within the Earth itself.

Thus the Moon's orbit is far more obvious and observable about the barycentre than that of the Earth: whilst the Moon orbits about the Earth at an average distance of 0.384×10^6 kilometres (0.239×10^6 miles) this is easily observable; the Earth in effect wobbles about its axis and in the same context this motion is far from discernable.

Free Fall

Isaac Newton compared the motion of the Moon to a cannon ball fired horizontally from a top of a high mountain. He imagined the mountain top to be above the Earth's atmosphere, so air resistance would not slow the motion of the cannon ball.

When the cannon ball was fired with a small horizontal velocity, it would follow a parabolic (curved) path and soon fall to the ground due to the effect of gravity. If more gunpowder were to be used the ball would be fired faster and its path would be less curved thus hitting the ground farther away.

If the cannon ball were fired fast enough, Newton reasoned, the projectile would travel so fast as to never hit the ground. The parabolic path would become a circle and the cannon ball would circle the Earth indefinitely – in other words, it would be in orbit.

Newton's first law of motion states that an object in motion tends to stay in motion in a straight line unless acted upon by some force. Thus the cannon ball in this example travels in a curve around the Earth only because Earth's gravity acts to pull it away from its straight-line motion.

The cannon ball is effectively falling around the Earth in the same manner as the Moon. The Earth's gravity pulls it toward its centre, but the Earth's surface curves away from it at the same rate at which it falls. Both the cannon ball and the Moon have 'sideways' velocity, or tangential velocity parallel to the Earth's surface, which is sufficient to ensure motion around the Earth rather than into it. If there is no resistance to reduce their speed, the Moon and cannon ball 'fall' around the Earth indefinitely.

Thus it should now be clear that when the astronauts in a NASA space shuttle are floating about in Earth orbit, they are not doing so because they are too far away to be affected by the Earth's gravity. It is because the downward acceleration of the spacecraft cancels the pull of the

Earth's gravity. <u>The orbiting spacecraft is falling and its orbital motion causes it to fall around the Earth the same as the cannon ball.</u>

How the Tides Operate

As mentioned at the beginning of this chapter, most people when asked how the ocean tides are raised would offer that it is mainly because of the pull of the Moon and to a lesser extent the pull of the Sun, implying that it is the effect of the direct gravitational attraction of these celestial bodies. In other words they are saying it is the Moon's gravity that is pulling directly on the Earth's oceans and drawing them upward in the direction of the Moon. Now this has to be questionable as the Earth attracts objects (masses) to its surface 287,956 times more than the Moon and 1,654 times more than the Sun!

If the tides were a consequence of purely the direct attraction of the gravitational pull of the Moon, then it could also be argued, theoretically, that the oceans would pile up on the side of the Earth nearest the Moon and there would not be an appreciable ocean bulge on the side of the Earth furthest from the Moon; indeed, ocean water on this side of the Earth would surely flow to replenish water drained from the Earth's surface at right angles to the Moon. Thus water would flow from the side of the Earth furthest from the Moon to replenish that drawn toward the Moon. Additionally when the Sun, Moon and Earth are in direct alignment with the Sun and Moon, on the same side of the Earth, then the combined effect of both the Sun and the Moon's gravitational pull would exacerbate the situation.

But we know there are other influences at work as the Moon orbits the Earth, and Earth with its Moon orbits the Sun; <u>it is these motions that are the key.</u> So if the ocean bulge is not as a consequence of purely the direct gravitational pull of the Moon and Sun, then how do these motions contribute to the tides we see on Earth; what forces are at work to cause these ocean tides as we know them?

With our understanding now of barycentre, free fall et cetera we shall now explore this phenomenon.

The gravitational forces of both the Sun and the Moon are obviously involved in the mechanics of the ocean tides, but not in the way most people perceive these forces to operate. Although the Sun has the greater mass – hence – the greater gravitational force – it is the Moon that has the most marked effect upon the ocean tides due to its closer proximity.

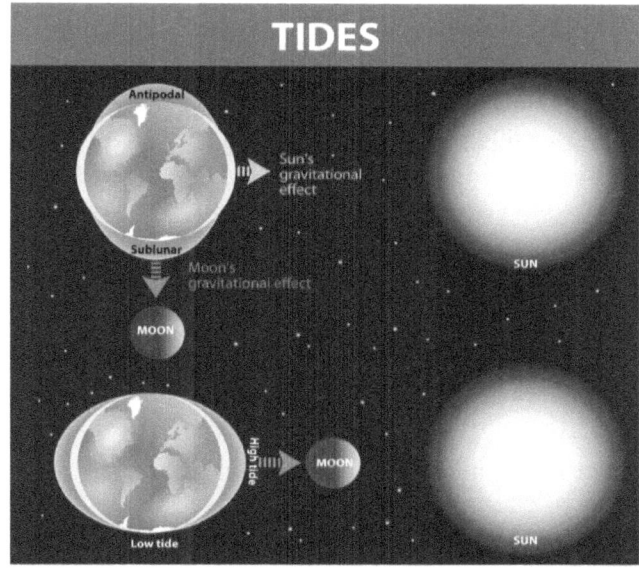

The effect of the gravitational force of one body upon another is dependent upon the inverse-square law (Isaac Newton).

Thus the Sun being some 400 times further from the Earth than is the Moon will have its effective gravitational force reduced much more substantially. Nonetheless as we have said it does have some effect in its own right upon the oceans of the world.

It should be noted that land tides also occur but they are on a very much smaller scale and as such they are not discernable to the human senses – although the probability of earthquakes and similar activity, is slightly higher when the planet is experiencing 'Earth spring tides' such as at new or full Moon. Further it should be noted there are atmospheric tides, but as the air has low mass these tides are not as pronounced as the land tides, indeed they are very small.

It is important to keep in mind what is happening to mass when it is in a state of Free Fall; remember that astronauts do not feel the pull of the Earth's gravity when orbiting the Earth in their space craft. Their weightlessness is not due to the reduction in the Earth's gravitational force, but due to the fact that the downward acceleration of their space craft exactly cancels the pull of the Earth's gravity – the astronauts are in Free Fall. Now it is also important to keep in mind that both the Earth and the Moon are in Free Fall around each other, or more precisely around their common centre (barycentre) of mass.

This means that similar to the case of the astronauts in Free Fall anything on the surface of the Earth or Moon will not experience the direct gravitational pull of each other. We also know the immensity of the Sun's gravity when compared to the Earth and the Moon. Indeed, it is this gravity that keeps the Earth in orbit around the Sun - but your weight is not affected when the Sun is overhead! Nor would your respective weight be affected by the Sun being overhead on Mercury and as we know Mercury is the closest planet to the Sun!

Tidal Gravity

We shall now see what we mean by tidal gravity and how Free Fall plays its part in the mechanism of the ocean tides.

The ocean tides are in fact due to the 'difference' in the way the Moon's gravity pulls on 'different' parts of the Earth as a result of the Earth falling toward the Moon; this difference in the Moon's gravity causes a stretching and squeezing of the oceans.

It is this stretching and squeezing that causes the Earth's tides. As we have already said the Earth and the Moon are in free fall around each other, or to be more precise, around their common centre (barycentre) of mass. This means that the oceans of the Earth are in effect falling toward the Moon and to see how this influences the oceans we need to carry out the following *thought experiment*:

Imagine a large container in the shape of a rectangular box about five thousand kilometres above the Earth's surface with two solid metal spherical objects inside placed on the floor – that is, the inside surface nearest the Earth's surface.

The large container with the two spherical objects is allowed to fall and as such will plummet towards the Earth's surface under the influence of gravity. It is important to remember that gravity operates from the geometric centre of the Earth and as such the gravitational field will have a radial configuration.

Thus any two falling objects will converge and (theoretically) meet at the centre of the Earth.

Now we know from Newton that if two such spherical balls were arranged adjacent to each other and about a metre apart in deep space with no other mass (or gravitational field) close by to influence the spheres, then the two balls would be attracted and move toward one another till they met. If both spheres were of identical mass they would move together at the same velocity, as they would be exerting the same force on each other.

Returning to our falling container with the two objects we can safely ignore the effect of their mutual gravity, as the Earth's gravity is so enormously much greater than that of the two spheres. Thus with the two objects falling toward the geometric centre of the Earth it means their paths will converge and as such they will move together.

Now that we have seen how two objects move together when in free fall, let us examine the case for four such spheres.

Imagine we now have four solid metallic and spherical objects falling toward the Earth. The objects are arranged to form a diamond shape with one point of the diamond (one sphere) nearest the Earth and leading the way down and the opposite point (top sphere) coming last.

All the four spheres are in free fall. The highest sphere is at the greatest distance from the Earth and as gravity decreases with distance, so the attraction between this highest object and the Earth is less than the attraction between its three companions and the Earth. Thus it lags further and further behind the others as they fall. The lowest sphere has the greatest attraction, as it is closest to the Earth and thus travels the fastest increasing its lead as the objects fall.

Four solid metal spheres fall toward the Earth in free fall. The differences in the gravitational pull of the Earth on each of them causes them to fall differently. The sphere nearest the Earth falls more rapidly than the others and gets ahead. The sphere farthest from the Earth falls less rapidly and lags behind.

The outside pair of spheres beside falling less rapidly than the sphere closest to the Earth also move together as shown in the diagram to the left.

The other two spheres fall at the same rate, but their paths converge. The ultimate result is that the diamond configuration of the four spheres becomes distorted and it is the difference in the strength of gravity from place to place that causes the distortion.

Now suppose we exchange the four metal spheres for one sphere made of soft plasticine and allow that to fall under the influence of gravity.

Thus the same forces will act upon the sphere of plasticine as in the previous case of the four metal spheres - as the plasticine sphere continues to fall the effects of squeezing and stretching the plasticine will cause it to become more ovoid in shape. This vertical stretch and lateral squeeze is due to gravitational forces known as tidal gravity.

From what we have learnt from the above *thought experiment* we can now go ahead in understanding the mechanism of the ocean tides.

The Moon and the Earth are in free fall orbit and nothing on the surface of the Moon or the Earth can feel the direct gravity of the other body. However, as happened in our thought experiment, the Earth, the Moon and things on the surface of them can be affected by the 'difference' in the strength of the gravitational pull from place to place and the 'difference' that makes in the way they fall.

The result being that the waters on the side of the Earth nearest the Moon bulge out moonward, making a high tide. The opposite side gets left behind a little, as in the case of the falling metal spheres, making a second high tide. In between there are low tides due to the squeezing effect and the 'squashed' waters contribute to the ocean bulges.

In explaining the mechanism of the ocean tides we have ignored the effect of the Sun's gravitational pull for although it does have an influence it is of a lesser nature due to its distance in comparison to the proximity of the Moon.

Although the Sun's mutual gravity is about 180 times stronger than the Moon's mutual gravity, solar tides are not as big as lunar tides. This is because of the difference between the Sun's gravity on the side of the Earth nearest the Sun and the side of the Earth furthest from the Sun is small. The Sun is very much further away from the Earth than the Moon and the distance from one side of the Earth to the other is so small in comparison to the distance of the Earth from the Sun that it makes very little difference in the effect imposed by the Sun.

To summarise our study of the ocean tides so far, we have stated that the bulge on the side of the Earth furthest from the Moon appears to be somewhat enigmatic when related to purely a direct gravitational force, as apart from any inertial effect one would expect most of the oceans to be pulled toward and bulge out on the surface of the Earth nearest to the Moon. If the oceans of the Earth's were subjected to the direct effect of the pull of the Moon then, theoretically, it could be argued that this would indeed be the result; the oceans would tend to pile up on the side of the Earth facing and nearest to the Moon. But we have seen this is not the case as the fundamental mechanism behind the ocean tides is due to the phenomenon known as tidal gravity.

Other Factors Contributing to the Ocean Tides

As the land masses, vis-à-vis the oceans, are not uniform there are many exceptions to the basic theory - there are negligible tides in the Baltic and the Mediterranean; being almost completely land locked there is no effective movement of water – as the water tends to move (under the effect of lunar and solar gravity) there is no water to flow into the resultant hollows and the back pressure prevents the forward movement - this of course particularly applies to lakes and totally inland seas. Due to its vast mass the Pacific Ocean does not follow the basic theory and in some parts has only one tidal cycle per day.

The difference between high and low tide can be considerable at places that are quite close together, as an example, on the French coast in Mont St Michel Bay the tidal range is approximately 14 metres whilst at Cherbourg, a hundred kilometres up the coast, it is only 6.6 metres. There are some even more startling differences in other parts of the World. In the Bay of Fundy (Canada) the maximum range is 19.69 metres, whilst at Portishead on the Bristol Channel it is 15.75 metres. At the other end of the scale in the mid-Pacific the tidal range is very low at just under 2 metres in the Marquesas. In the Mediterranean it is even lower and on the shores of the Adriatic it is only 1.3 metres.

There are also periodic variations in the interplay of the two prime gravitational forces that affect the tidal heights caused by the elliptical orbit of the Earth around the Sun and that of the Moon around the Earth.

The study and understanding of ocean movements and the theory of tides is a very complex subject. The variation of the seashores, channels, estuaries, steepness of beaches, ocean currents and size of the ocean all play their part. As we have seen above there is a very marked difference in tides of the Mediterranean and that of Mont St Michel Bay, Northern France, Therefore the ultimate and actual behaviour of the tides lies not in the province of astronomy but in that of hydrography and physical geography.

Spring Tides

The highest tides, known as Spring Tides, occur around the time of a New and Full Moon. That is when the Sun, Moon and Earth are in direct alignment.

Spring Tides result in higher-than-average high tides and lower-than-average low tides. A consequence of the elliptical nature of the orbits of the Earth and the Moon is that the distance between the Sun, Earth and Moon vary. This means that all Spring Tides are not equally high. Indeed, the distance between the Earth and the Moon varies by about 10 per cent, which results in a 30 per cent difference in the Moon's effect in raising tides.

Neap Tides

The Sun and Moon's tidal pull are virtually at right angles during the first and last quarter of the Moon. As such the ocean tides have their lowest amplitude.

These tides are known as Neap Tides.

Therefore the high tides are lower than average and the low tides are not as low as the average low tides.

Summary

In summary, the reason behind the ocean tides can be stated as:

The result of the Earth falling toward the Moon, and to a lesser extent, the Earth and Moon system, falling toward the Sun.

This being dependent on:

- ❖ The mass (gravity) of the Moon.

- ❖ The mass (gravity) of the Sun.

- ❖ Rotation of the Earth on its axis.

- ❖ Position of the Earth, Moon and Sun with respect to one another.

- ❖ Varying distance between the Earth and the Moon.

- ❖ Varying distance between the Earth and the Sun.

- ❖ Inclination of the Moon's orbit.

- ❖ Variation in the shape of coastlines and relief of ocean basins.

There are generally two high tides and two low tides every 24 hours due mainly to the Moon's gravitational attraction and the motion of the Moon and Earth.

Finally, having dealt with the basic mechanisms of ocean tides it is deemed outside the scope of this book to address the subject of terrestrial tides any further. As mentioned earlier, the study and understanding of ocean movements and the theory of tides is a very complex subject. Therefore those wishing to pursue the matter further should be aware there are many books encompassing hydrography and physical geography on the market to adequately satisfy this need.

"Everyone acquainted with the subject will recognise it as a conspicuous failure."

Henry Morton, president of the Sevens Institute of Technology, on Edison's light bulb, 1880

CHAPTER FIVE

Energy Alternatives

We are all familiar with the saying that you cannot make an omelette without cracking eggs - the sentiment can equally be said to apply to the electrical energy producing industry - if we want the continuing luxury and the comfort of such things as electrical lighting - heating - appliances - computing - communication and entertainment in their many forms, et cetera, then there is going to be a price to pay for it – indeed there is no such thing as a free lunch as they say - the 'trick' is in keeping the cost as low as possible, coupled with minimum impact on the environment, and achieving sustainability.

We know only too well that the burning of fossil fuels, such as coal, emit atmospheric emissions and supplies are finite - nuclear fission rules itself out on the grounds of its horrendous cost, hazardous waste and the possibility of an Armageddon scale disaster - wind generation offers little but a limited and sporadic output for the industrialisation and desecration of acres and acres of beautiful countryside - no one in their right mind would be considering the proliferation of wind generators across the UK without the government subsidy.

Solar Radiation

I find it incredible that oil-rich Middle Eastern countries are not spending a large portion of their oil revenues preparing for the days when the oil runs out, as they surely will. Indeed, countries along the coast of North Africa could have vast solar arrays feeding electrical energy, not only to their own lands, but Europe as well using High Voltage Direct Current (HVDC) 'interconnectors', across the Straits of Gibraltar, and from the North African coast via Malta and Sicily to Italy. When consideration is given to the 'solar radiation rich' areas of Africa, such as the Sahara it would appear the exploitation of the Sun to be endless - supplying electrical energy to many countries – almost in the sense of the oil exporting countries, except HVDC lines would obviously replace ships.

The same could be said for the other solar radiation rich countries across the planet – additionally, solar radiation countries that have a coast line could run electrically driven desalination plants, and therefore have an abundance of usable water – offering the opportunity of irrigating sunny, but parched lands – creating 'Gardens of Eden' all over the place - all it takes is the political and industrial will.

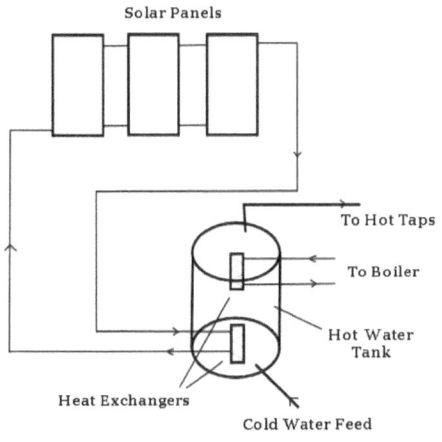

The diagram shows a typical solar hot water heating system. Solar heated water in the solar panels is fed, either by convection or pump, to a heat exchanger in the hot water tank. When the sun is not shining, hot water is obtained from a separate boiler via its own heat exchanger.

Unfortunately for the UK large solar arrays are not deemed viable due to lesser solar radiation at our latitudes offering limited and lower power outputs - especially during the winter months. Nevertheless solar panels can make a marked difference in energy savings at homes, in offices and industrial sites in the UK - small solar arrays on the roofs of domestic properties will prove economically rewarding – more so with the government subsidies available, see Chapter Seven. It was Thomas Edison, USA, who said, "I'd put my money on the Sun and solar energy. What a source of power! I hope we don't have to wait until oil and coal run out before we tackle that."

There are two types of solar panel, one that heats water and the other that generates electricity. Solar water heating systems use the solar panels to heat water by gathering energy radiated from the Sun; solar water heating systems work in conjunction with conventional water heating systems. Thus a typical solar water heating system for domestic hot water consists of three main components, namely, solar panels, a hot water cylinder and the necessary plumbing system; the solar panels are usually fitted to a predominantly south facing roof of the property, whereby the solar collectors within the panels will absorb and retain heat from the Sun's ray, transferring the heat energy to the water inside the collectors.

This hot water is then, depending on the system, fed by convection or pump, to a heat exchanger contained within a hot water cylinder. Heat is then transferred from the heat exchanger to the water within the hot water tank. The hot water is stored in the cylinder until it is required. A second heat exchanger is usually found within the same hot water cylinder, but in this case it is plumbed to the house main central heating and hot water boiler. This arrangement ensures that hot water is available both day and night, but also when the Sun is not shining. The solar water heating system obviously does not produce any electricity, but saves electricity and other forms of energy indirectly by using the heat from the Sun.

The other type of solar panel is made up of a number of Photovoltaic (PV) cells. This type of system use cells to convert solar radiation into electricity. It should be noted that PV cells will produce electricity in daylight as well as direct sunlight, although they operate most effectively in direct sunlight. The phenomenon whereby certain materials would produce small amounts of electrical energy when exposed to light goes back to the 19th century. Although the photovoltaic cell as we know it nowadays, is the result of a photovoltaic module developed by Bell Laboratories in the United States. The first cells were used to provide power abroad spacecraft and it was through the various space programs that the technology

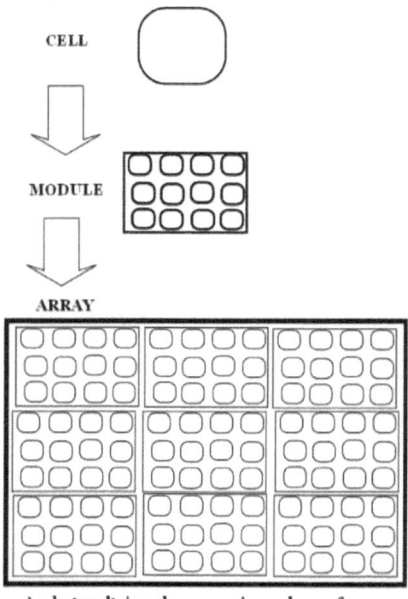

A photovoltaic solar array is made up from a number of modules. The module, in turn, is made up from number of PV cells.

advanced to its current effectiveness and reliability. Indeed, the Hubble Space Telescope, for example, employs photovoltaic solar panels for its electrical energy requirements. The PV cell consists of one or two layers of a semi-conducting material, such as silicon as used in the microelectronics industry.

Thus when light energy falls upon the cell it creates an electric field across the layers, causing electricity to flow i.e. electrons are knocked loose from the atoms in the semiconductor material, so if electrical conductors are attached to the positive and negative sides of the semiconductor and attached to a suitable load, then a current will flow;

the greater the intensity of the light, the greater the current flow. A number of photovoltaic cells mounted together, and connected electrically to each other in a suitable support structure or frame, is called a photovoltaic module. Where a number of modules are wired together in a larger arrangement this is termed an array. It should be noted that photovoltaic modules and arrays produce direct current and as such can be connected in both series and parallel to produce any required voltage and current arrangement.

There are obviously many applications for PV cells such as calculators, battery chargers to providing homes, offices and factories with electrical power. One advantage of using this type of solar array is that any excess power can be fed (sold) into the local electricity system - thus homeowners when away on their summer holiday can relax in the knowledge that their array of solar panels are producing (environmentally friendly) and feeding (selling) power to their electricity supplier.

Tidal Energy

Regarding the United Kingdom and that of renewables, part of the solution is staring us in the face – especially those who live on the coast - it also makes us unique from the rest of Europe and many other places - we are not just an island, but fortunately, an island which experiences *high tidal energy* - in fact the Bristol Channel has the second highest rise and fall of tide in the world. So why are we wasting millions and millions of pounds on limited and costly onshore and offshore wind technology, and billions of pounds on nuclear fission - it is as though the inmates have taken over the asylum. We should be taking advantage of the energy found in the oceans such as tidal, wave and ocean currents - the technology whereby electrical energy may be obtained via the power of the oceans encompassing estuary and ocean impoundments, tidal fence turbines, tidal current generators and wave technology.

It truly beggars belief that tidal, wave and ocean current energy do not dominate the list for a sustainable, clean source of energy for the production of electrical energy for our islands. The United Kingdom has a land mass of 244,690 square kilometres (94475 square miles), of which England is 130,328 square kilometres (50,320 square miles) and Wales 20,759 square kilometres (8,015 square miles) and a very long coast line at over 11,000 miles in length – remember the old adage of 'not being able to see the woods for the trees', but as we shall see later in this chapter, perhaps the tide is now, at last, beginning to turn (pun intended).

Regarding tidal energy, as has already been pointed out, we have the second highest rise and fall of tide in the world - this of course is the Bristol Channel. Thus is it beyond the wit and brilliance of British skills and engineering not to be able to harness all this energy - it is useful to remember that it was the British architect, Norman Foster, whose design was chosen for the 'Viaduc de Millau' in France; this viaduct is 1.5 miles long, and at the time of writing, is the tallest road bridge in the world. Some people thought it could not be done - but it has - so much for their faith and vision.

It was not so long ago that an Astronomer Royal stated quite categorically, that space flight was utter bilge - so go tell Neil Armstrong - and it was certainly a good thing that the Wright Brothers ignored their detractors, and the then 'obvious notion' that heavier than air machines could not fly - a pity these critics are not around today for a flight on a Jumbo Jet!

Indeed, none other than the French Academy of Science stated that meteorites did not fall to Earth and the phenomenon did not exist - although we now know different – and it is only relatively recently that scientists have accepted the phenomenon of ball lightning – and no doubt, dear reader, you can think of many more examples.

What we urgently need now is positive thinking and the moral courage to achieve our goals. Given the necessary government backing and the subsidies being afforded to the wind farm industry, especially the nuclear power industry, then of course we can 'tap' ocean power. You only have to study history to see that given the will and determination, British scientists, engineers and architects are amongst the best in the world.

We once led the world in scientific discovery – recollect who invented the electrical generator and the steam turbine – it was Michael Faraday and Charles Parsons respectively. All the following were British inventions: steam engine (Thomas Newcombe, 1712), improved steam engine (James Watt, 1765), steam locomotive (George Stephenson, 1814), military tank (Ernest Swinton, 1914), reflecting telescope (Isaac Newton, 1668), Terelyne (John Whinfield, J.T. Dickson, 1941), flushing toilet (Joseph Bramah, 1778), stainless steel (Harry Brearley, 1913), spinning frame (Richard Arkwright, 1769), shrapnel shell (Henry Shrapnel, 1784), seismograph (John Milne, 1880), seed drill (Jethro Tull, 1701), steam road locomotive (Richard Trevithick, 1801), achromatic lens (Chester Moor Hall, 1733), carbon fibre (Leslie Phillips, 1963), Portland Cement (Joseph Aspidin, 1824), early computer (Charles

Babbage, 1822), diode valve (Ambrose Flemming, 1904) and so on, the list is endless and again dear reader you can think of many others.

Then what of the great discoveries from Newton to Hawkins, again we could create an endless list - what of our great architects and civil engineers – so it truly does beggar belief, that with our inherent skills, there are those myopic people who would deny us from tapping the seas around our coast for environmentally friendly electrical energy - remembering that, as stated at the beginning of this chapter, nothing is totally environmentally friendly – there is always a price to pay in this respect.

The tidal race that sweeps through the Menai Straights is begging to be tapped - any person who has sailed through these straights will tell you of their respect for these waters. The Menai suspension bridge (579ft-long) built in 1826 by Thomas Telford was the first of its structure in the world rising above the dangerous rip tides that made a ferry journey a dangerous option earning the Menai the nickname, 'the British Bosphorous'.

On a smaller scale we have the tidal race between Ramsey Island and the Pembrokeshire coast line. Remember, unless the Moon decides to depart from its orbital path, this tidal power is available both night and day - a continuous and clean source of energy – with high and low tide occurring at different times along the coast.

According to the World Energy Council (WEC) and with their kind consent, I quote from their website at www.worldenergy.org as follows:

"Tides are caused by the gravitational force of the Moon and the Sun acting upon the oceans of the rotating Earth. The relative motions of these bodies cause the surface of the oceans to be raised and lowered periodically, according to a number of interacting cycles. These include:

- ❖ A half day cycle, due to the rotation of the Earth within the gravitational field of the Moon.

- ❖ A 14 day cycle, resulting from the gravitational field of the Moon combining with that of the Sun to give alternating spring (maximum) and neap (minimum) tides.

- ❖ A half year cycle, due to the inclination of the Moon's orbit to that of the Earth, giving rise to maxima in the spring tides in March and September.

- ❖ Other cycles. Such as those over 19 years and 1,600 years, arising from further complex gravitational interactions.

The range of a spring tide is commonly about twice that of a neap tide, whereas the longer period cycles impose smaller perturbations. In the open ocean, the maximum amplitude of the tides is about one metre. Tidal amplitudes are increased substantially towards the coast, particularly in estuaries. This is mainly caused by shelving of the sea bed and funnelling of the water by estuaries. In some cases the tidal range can be further amplified by reflection of the tidal wave by the coastline or resonance. This is a special effect that occurs in long, trumpet-shaped estuaries, when the length of the estuary is close to one quarter of the tidal wave length.

These effects combine to give a mean spring tidal range of over 11 metres in the Severn Estuary (UK). As a result of these various factors, the tidal range can vary substantially between different points on a coastline. The amount of energy obtainable from a tidal energy scheme therefore varies with location and time.

Output changes as the tide ebbs and floods each day; it can also vary by a factor of about four over a spring-neap cycle. Tidal energy is, however, highly predictable in both amount and timing. The available energy is approximately proportional to the square of the tidal range. Extraction of energy from the tides is considered to be practical only at those sites where the energy is concentrated in the form of large tides and the geography provides suitable sites for tidal plant construction.

Such sites are not commonplace but a considerable number have been identified in the UK, France, Eastern Canada, and Pacific coast of Russia, Korea, China, Mexico and Chile. Other sites have been identified along the Patagonian coast of Argentina, Western Australia and Western India.

Tidal energy can also be exploited directly from marine currents induced by the combined lunar and solar gravitational forces responsible for tides. These forces cause semi-diurnal movement in water in shallow seas, particularly where the coastal morphology creates natural constrictions, for example around headlands or between islands. This phenomenon produces strong currents, or tidal streams, which are prevalent around the

British Isles and many other parts of the world where there are similar conditions. These currents are particularly prevalent where there is a time difference in tidal cycles between two sections of coastal sea. The flow is cyclical, increasing in velocity and then decreasing before switching to the opposite direction. The kinetic energy within these currents could be converted to electricity, by placing free standing turbo-generating equipment in offshore areas.

Different Technical Concepts for Exploiting Tidal Energy

Most countries which have investigated the potential exploitation of tidal energy have concentrated on the use of barrages to create artificial impoundments that can be used to control the natural tidal flow. Barrage developers in the UK and elsewhere concluded that building a permeable barrage across an estuary minimises the cost of civil structures for the quantity of energy that can be realistically extracted.

Construction of barrages across estuaries with high tidal ranges would be challenging but technically feasible. In shallow water armoured embankment would be used, but in deeper water this method would be impractical and too expensive because of the quantity of material required. Complete closure of estuaries would be achieved by emplacing a series of prefabricated sections, or caissons, made from concrete or steel which could be floated and then sunk into position.

The technique has been used in the Netherlands to close the Schelde Estuary. A large steel caisson was used in the construction of the Vadalia power station on a tributary of the Mississippi. Tidal barrages would comprise sluice gates and turbine generators. Large scale structures like the Severn Barrage would also include blank caissons and ship-locks. During the ebb tide water is allowed to flow through the sluices and the turbine draft tubes to ensure the maximum possible passage of water into the impounded basin. At or close to high water the sluice gates are closed. At this stage of the cycle the turbines can be used in reverse as pumps to increase the amount of water within the basin. Although there is an obvious energy demand, the amount of water transferred can provide an additional increase in energy output of up to 10 per cent compared with a cycle where no pumping is used. The actual increase in energy output from pumping depends on the estuary and the tidal conditions. Retention of water allows a head of water (i.e. difference in vertical height of water levels) to be created as the flood tide progresses seaward of the barrage is no different to a low-head hydro-electric dam. The large volumes of water and the variation in head require the use of double regulation, or Kaplan turbines.

These turbines have guide vanes and blades that can be moved by hydraulic motors. This allows turbine operation, and therefore energy conversion efficiency, to be optimised through each generation cycle as the reservoir head drops.

Experience from the UK's tidal energy programme revealed that ebb generation (i.e. only on the ebb tide) maximises the amount of energy that can be produced from this type of barrage system. Two-way generation (on both the flood and ebb tides) is technically possible, however less energy would be produced because the head of water created prior to generation is lower compared with an ebb generation cycle. Moreover, Kaplin turbines in a horizontal configuration are optimised for generation with the flow in one direction.

As with all other civil engineering and power generation projects, diligent technical appraisal is essential to mitigate against both technical and commercial risk. Barrage design requires a detailed geotechnical site investigation to determine the foundation conditions. The nature of the substrate and the dimensions of an estuary ultimately determined the design options for barrages. Once an optimal design has been identified, it needs to be developed in detail to establish the construction schedule and the costs at each stage of the project to determine both economic and financial viability.

A detailed knowledge of the hydraulic flow pattern before and after the barrage has been constructed is of equal importance and for the same reason. Hydraulic flow has to be accurately modelled, using complex mathematical models that can accurately simulate natural flow conditions, so that the effects of progressive closure and environmental changes can be predicted. Hydraulic modelling is also used to determine the energy output from the system during each tidal cycle. Other concepts based on secondary artificial storage systems have been investigated, and continue to be promoted. The concept enables storage within two or more basins which can increase the control of the water movement and allows the turbines to operate for longer than in single basin schemes. Secondary reservoirs were proposed for the Severn scheme but were discontinued because of the cost of the energy produced. The rise in cost is the direct consequence of the substantial additional civil structures required.

Technical Status and Experience from Operating Systems

Tide mills were commonplace along the coasts of Western Europe from the Middle Age, until the Industrial Revolution supplanted renewable forms of energy with fossil fuel alternatives. Interest in tidal energy was stimulated by the construction of the French barrage across the Rance estuary in Brittany during the 1960's. A dam was built in-situ between two coffer dams. Consequently the entrapped estuarial waters stagnated, although the ecosystem recovered once the barrage began operation. Most of the structure, which has an installed capacity of 240 MW, is comprised of Kaplan turbines with only a small bank of sluices. The barrage has a ship lock adjacent to the control centre and carries a trunk road. Originally designed for two-way generation, the operators, EDF, predominantly generate on ebb tides. Despite over thirty years of successful operation, EDF have no plans to build other barrage schemes.

Shortly after the completion of the Rance barrage, the Russians built a small experimental system with an installed capacity of 400 kW. The scheme was constructed at Kislogubsk near Murmansk, partly to demonstrate the use of caissons in barrage construction.

The potential for tidal energy at the head of the Bay of Fundy, which extends between the Canadian maritime provinces of Nova Scotia and New Brunswick, has long been recognised. In 1984 a 20 MW plant was commissioned at Annapolis, across a small inlet on the Bay of Fundy's east coast. The barrage was built to demonstrate a large diameter rim-generator (Straflo) turbine. Despite the large tidal energy potential, Canada has relied upon the development of its substantial conventional hydropower reserves.

The UK has invested approximately £20 million in tidal energy Research and Development (R&D). Most of this effort was concentrated on co-funded feasibility and development studies (between the mid-1980's and 1992) with industrial consortia. Two main sites were evaluated: one on the Severn (mean spring tidal range 12 m); and the other on the Mersey Estuary (mean spring tidal range 8 m). Despite detailed technical appraisals, coupled with evaluations of the effects to shipping and the environment, neither project progressed beyond an early development stage. The work revealed that tidal energy was less economic compared with other forms of renewable energy. The UK Programme also investigated four smaller-scale projects (ranging in size from 5-100 MW). None of these schemes progressed further than initial feasibility.

The large tidal range along the west coasts of England and Wales provides some of the most favourable conditions in the world for the utilisation of tidal power. If all reasonably exploitable estuaries were utilised, annual generation of electricity from tidal power plants would be some 50 TWh, equivalent to about 15 per cent of current UK electricity consumption - of six identified sites with mean tidal ranges of 5.2-7.0 m. feasibility studies have been completed for two large scheme: Severn estuary (8,640 MW) and Mersey estuary (700 MW) and for smaller schemes on the estuaries of the Duddon (100 MW), Wyre (64 MW), Conwy (33 MW) and Loughor (5 MW).

A governmental programme on tidal energy (1978-1994) concluded that given the combination of high capital costs, lengthy construction periods and relatively low load factor (21-24 per cent), none of these schemes is regarded as financially attractive in present circumstances. A future UK tidal energy programme could include construction of a small-scale scheme primarily to demonstrate the technology and its environmental effects, before progressing to very large schemes on the scale of the Severn." End of quote with acknowledgements to WEC.

Now it does make you think when a UK governmental programme on tidal energy (1978-1994) concludes that because of high capital costs, lengthy construction periods and relatively low load factors, they deem none of the UK schemes currently attractive?

Yet government is quite happy to industrialise and desecrate acres and acres of beautiful countryside with subsidised, ineffective and costly wind generators, which due to the very nature of the wind have low load factors. According to this governmental programme, tidal schemes have low load factors, which happen to be of the same order as wind generators – so why wind generation which is unpredictable and thus very difficult for grid control - compounding this enigmatic governmental approach is the fact that Scottish wind farms are being paid millions of pounds for NOT producing electricity when they have to shut down due to strong winds, such is the madness of it all - the tides FLOOD and EBB for 24 hours each day, and most importantly, are RELIABLE - not like the unreliable wind - it can be argued that tidal easily wins as a good, intelligent choice – so why is the UK investing in wind generation, it just does not make any sense.

Further, what are we to conclude when government uses as an argument, high capital costs and lengthy construction periods - of course there are high capital costs as in any large enterprise and of course there will be lengthy construction periods - what else do they expect - you cannot build any large power station overnight, a large conventional fossil fuelled power station can take 10 years from planning to commissioning - so what are they talking about - it is the long term benefit that should predominate and not the inadequate and ineffective short-term fixes such as wind, which at the end of the day we all pay dearly for.

Ocean Impoundment

According to Tidal Electric, a Connecticut company with offices in London, the paradigm for tidal power has been the tidal barrage. Although it has been in use for more than 1000 years, the tidal barrage is unsuitable for broad-scale commercial use because of environmental and economic drawbacks due, primarily, to its shoreline location. Offshore tidal power generation utilises an offshore impoundment structure (known as tidal lagoons) built of rubble mound construction materials (loose rock, sand, and gravel) sited in a shallow tidal flat with a large tidal range.

Placing the impoundment structure offshore resolves the environmental and economic problems of the tidal barrage and reintroduces the vast potential of the oceans' tides to the array of generation choices at the dawn of an era in which renewable source power is evolving from a marginal to a mainstream technology choice.

Offshore tidal power generators use familiar and reliable low-head hydroelectric generating equipment, conventional marine construction techniques, and standard power transmission methods.

At the time of writing Tidal Electric have three projects (Swansea Bay 320 MW, Fifoots Point 320 MW, and North Wales 432 MW) in development in Wales where tidal ranges are high, renewable source power is a strong public priority, and the electricity market place gives it a competitive edge.

Thus rather than blocking an estuary with a barrage, offshore tidal power generators use an impoundment structure, making it completely self-contained and independent of the shoreline (visualise a circular dam, built on the seabed), thereby eliminating the environmental problems associated with blocking off and changing the shoreline. Migratory fish simply swim around the structure, whereby ships and boats navigate past the structure. The optimal size for offshore tidal power generation is the shallow water of near-shore areas, while shipping lanes require deeper water.

The offshore siting is the distinctive characteristic of the design and one of the fundamental claims of its patents. Turbines are situated in a powerhouse that is contained in the impoundment structure and is always under water. Power is transmitted to shore via underground/underwater cables and connected to the grid. The structure need not be more than a few yards beyond the low tide level and the optimal site is one that is as shallow as possible, thereby minimising the cost of building the impoundment wall.

Tidal barrages must generate primarily in one direction (on the ebb tide) in order to minimise progressive disruption of the intertidal zone that would eventually lead to the silting up of the head pond. The offshore tidal power generator is free to utilise both the ebb and the flood tides for generation, thereby roughly doubling the load factor of the barrage. Double the load factor is equivalent to halving the capital cost per unit output. Additionally, multi-cell impoundment structures provide higher load factors (about 62 per cent) and have the flexibility to shape the output curve in order to dispatch power in response to demand price signals. Both the impoundment structure and the barrage are intended to hold back water.

The power of the tides lies only in the tidal range, that is, the difference in water levels between high tide and low tide. The impoundment structure is built so as to perform only that function. Whereas the barrage also holds back all the water below low water level and all the water in the intertidal zone. None of this water produces any power, yet it is very costly to contain. Additionally, the barrage obliges the out-migrating and returning fish to pass through the turbine, whereas the offshore impoundment structure presents no more of a hazard to fish than would a new sandbar and fish instinctively avoid swimming through a turbine.

The offshore tidal generator uses conventional low-head hydroelectric generation equipment and control systems. The equipment consists of a mixed-flow reversible bulb turbine, a generator, and the control system. Low-head hydroelectric generation equipment has been in existence for more than 120 years and state-of-the-art equipment is mature, mechanically efficient (96+ per cent), familiar (over 100,000 units in use world-wide), reliable and durable (the equipment comes with performance guarantees and a design life of over 50 years).

Manufacturers/suppliers include Alstom, GE, Kvaerner, Siemens, Voith, Sulzer and others.

Again we have to question the wisdom of government subsidies and the encouragement to the proliferation of wind farms across the United Kingdom. We know only too well, that wind generators are totally at the mercy of the wind and are revealing load factors of 25 per cent and less. Whereas, according to Tidal Electric, tidal lagoons can show load factors of 62 per cent - wind farms desecrate the countryside whilst the employment of tidal lagoons can be claimed to be very environmentally friendly - no contest I would suggest.

Tidal Fence

This type of technology calls for a tidal fence to be built between two areas of land, with a series of turbines contained in large caged concrete structures which turn slowly at 25 revolutions per minute (rpm), thereby allowing fish to swim through. A suitable gap is left in the fence to enable larger marine life such as seals and whales to swim through during times of slack tide or reduced flow.

Blue Energy Canada Inc. is a Canadian technology company committed to developing large-scale renewable ocean energy for the 21st Century. Their tidal fence power system is comprised of multiple vertical axis Davis hydro turbines, which convert the kinetic energy of ocean currents and tides into, clean, low-cost and sustainable electric power. A tidal array can be installed across an estuary with depths of 65 metres of less and tides of about 1 metre or more, across a passage where tidal velocities exceed 2 metres per second. Models range from 4 kW river models to 14 MW large ocean class models.

Low capital cost, thin shell marine caissons can be linked together to form a tidal fence, which is substantially less costly than large ocean dams and barrages. Blue Energy Canada Inc. claim that with sea water being 832 times as dense as air, thereby offering high energy density, their turbines can provide peak power outputs in the multiple Gigawatt level. A typical installation across a 1 kilometre crossing can produce more electrical energy than a large atomic energy power plant– cleanly, safely and cost effective.

Their proposed four kilometre tidal fence in the Philippines is estimated to produce some 2200 MW of peak power (1100 MW base average power). The technology offering minimal environmental impact as the open sluice design and slow rotation of the turbines (25 rpm) allows for virtually unimpeded fish and silt migration; producing zero greenhouse gas emissions and earning large amounts of tradable emission credits.

They also claim multiple infrastructure development with single infrastructure cost as the tidal fence can also serve as a transportation corridor, carry water mains, telephone and power lines bringing significant economic and social value.

Ocean Thermal Energy Conversion

Although perhaps not so significant to British coastal waters, but there are often significant temperature difference in the world's tropical oceans between the warm surface waters and the colder deep waters. Using an evaporator to drive a turbine, electrical energy could be generated when there is a difference of more than 20 degrees Celsius between the warmer and colder waters. This technology is still in its infancy and there is a long way to go before it becomes commercially viable.

Wave Power

Another source of energy from the oceans has to be the waves, as apart from the tides, ocean waves themselves offer an environmental friendly means of producing electrical energy. Indeed, the energy from waves alone could supply all of the world's electrical energy needs many times over. Wavegen are a marine renewable energy company, with its head office in Inverness, Scotland. They claim they are a world leader in wave energy and wave power. They say their marine renewable wave power specialists offer a unique blend of engineering, scientific and commercial skills. They state they are experts in wave energy technology, project development and management, as well as current and potential applications of wave energy. Wavegen is owned by Voith Siemens Hydro Power Generation which is part of the Voith Group. For further information visit: (www.voithsiemens.com).

Wavegen have developed LIMPET, the world's first commercial-scale wave energy device that generates wave energy for the grid. The LIMPET unit on the Island of Islay, off the west coast of Scotland has an inclined Oscillating Water Column (OWC) that couples with the surge-dominated wave field adjacent to the shore. The water depth at the entrance to the OWC is typically seven metres. The design of the air chamber is important to maximise the capture of wave energy and conversion to pneumatic power. The turbines are carefully matched to the air chamber to maximise power output. The performance has been optimised for annual average wave intensities of between 15 and 25 kW/m. The water column feeds a pair of counter-rotating turbines, each of which drives a 250 kW generator, giving a nameplate rating of 500 kW. The unit is designed to operate, apart from flat calm, in all weathers, including storm conditions.

Cliff-Based Wave Power Potential for UK

Wave power plants built into cliff faces around the UK coast could be a reality within the next decade thanks to a pioneering new wave energy plant being developed in a cliff on the Faroe Islands. The Inverness based company Wavegen have also been working in partnership with SEV, the Faroes electricity company, to develop a wave power station in the Faroe Islands.

The Faroes wave power station is based on a series of Wavegen's air turbine power generation modules. The entire project, worth up to £7 million, is a blueprint for wave power stations in similar locations both in the Faroes and other parts of the world; the company say the joint venture brings together Wavegen's world-class experience in harnessing wave energy and the tunnelling experience of the Faroes. The Faroes power station is based on the Oscillating Water Column technology successfully developed by Wavegen at its Islay plant. The key innovative feature is the use of tunnels cut into the cliffs on the shoreline to form the chamber which captures the energy. The new design offers a novel and complimentary approach to shoreline devices that is well-protected and unobtrusive. It also overcomes one of the main challenges facing the development of onshore wave power in the Faroes – the high cliffs that surround the islands.

Tom Heath BSc, PhD, M.I.Mech.E C.Eng, Engineering Manager at Wavegen in his discussion titled, 'Realities of Wave Technology', has this to say and I quote with the kind consent of Wavegen, "Just as electrical power is not stored in transmission lines, but flows from the generator to the user, so wave energy is not static, but flows in the direction of wave propagation. If the energy flowing past a particular point in the ocean or arriving at a shoreline is not captured there then it is lost. Fortunately for the wave energy industry if one wave is lost there will be another one along soon bearing more energy. Because the energy is flowing we can consider the amount of power in kW contained in each linear metre of wave front. Off the west coast of Scotland the ocean has some of the best wave energy on the planet - but there is a wide variation between the differing power available in different seasons of the year. Off the West coast of Scotland the winter availability may be four times the summer average.

In our part of the world this can be considered an advantage because our energy demand in the cold season is so much higher than that in the summer months. This is not necessarily true worldwide. The Atlantic seaboard of the British Isles has one of the best wave energy climates on the planet with 60-70 kW/m (power flux) in deep water off the Western Isles falling to 15-20 kW at the shoreline as the effects of bottom friction and wave breaking take their toll. With the land mass of Scotland offering shelter to the south west the available power falls as we move east along the northern coast of Scotland but is still a remarkably attractive 25-50 kW/m (dependent on water depth) by the time we reach waters of Orcadia. The reality is that the power is there, the challenge is to harness it." End of quote.

Ocean Currents

A completely new kind of energy system, which uses the almost limitless energy of flowing sea currents, has been successfully, installed approximately 3 kilometres to the NE of Lynmouth in North Devon.

Marine Current Turbines (MCT) herald the enterprise as a 'World First' being the most powerful device of its kind so far installed with a rated power of 300 kW – making it potentially capable of meeting the average electricity needs of about 200 typical UK households.

It is also the world's first marine renewable energy system of significant size to be installed in a genuinely offshore location as previous marine renewable energy systems, whether for tidal or wave energy, have been located onshore or in sheltered, largely land-locked waters. This project marks the stage at which the technology for exploiting marine energy has moved

Artist impression of MCT's SeaGen device.

(Reproduced by kind permission of MCT.)

for the first time into the harsher energy-rich environment in which it needs to operate.

The turbine is culmination of the 'Seaflow' project, a £3.5 million project that is being conducted by an industrial consortium of UK and German companies and supported by the UK Department of Trade and Industry (DTI), the Joule Programme of the European Commission, and German Government. The project is aimed at testing the prototype turbine, and demonstrating technology, which will be further developed to a commercially viable stage by Marine Current Turbines Ltd over the course of the next few years.

The Seaflow Project represents the first phase of a comprehensive R&D programme intended to develop pioneering technology for exploiting the energy of marine tidal currents. The technology consists of underwater propellers mounted on steel piles (tubular steel columns) set into a socket drilled in the seabed. The propellers are driven by the flow of water in much the same way that wind generator propellers are driven by the wind, the main difference being that water is more than 800 times as dense as air, so quite slow velocities in water will generate significant amounts of power.

This project, in effect, involves the development of an 'Underwater Windmill' which can generate a maximum of 300 kW in a 2.7 m/s current (5.5 knots). The energy generated, being derived from tides has the significant advantage of being predictable. Maintenance of the device while it is submerged in fast currents would be exceptionally challenging and expensive, so a key patented feature of the technology is that the rotor and drive train (i.e. gearbox and generator) can be raised completely above the surface. Once raised, any maintenance or repairs can readily be carried out from the structure attended by a surface vessel. The project involves the design, manufacture, installation, testing and demonstration of the turbine, which will provide the essential information needed to design and build larger systems for commercial power generation, which will follow during the next few years.

The prototype experimental unit was successfully located 1 km off Foreland Point (approximately 3 km NE of Lynmouth, Devon, UK) on 26th May 2003. The territorial waters off Wales offer great potential for tidal stream technology, and an opportunity exists for well-sited demonstration projects to provide impetus to this new and emerging form of generation.

Marine Current Turbines plan to identify sites for potential development and ultimately progress at least one of these sites to install a 10 MW tidal stream farm (an array). A 10 MW tidal stream array would generate sufficient energy to meet the equivalent demand of approximately 7,000 homes. But the subsequent development of the array will be dependent on the outcome of a detailed environmental impact assessment, engineering studies, as well as consultations with local and national interests. The installation of the array will also require the necessary planning consents as well as further financial investment. Just imagine if all the effort and huge subsidies that have gone into nuclear technology over the last fifty years had been available for ocean energy technology - I very much doubt if we would have an electrical energy problem of the same magnitude as we do today.

Hydro-Electricity

As we have seen earlier in this book hydro-electricity was a dominant force in the evolution of the electrical generating industry in the USA. In a hydro-electrically operated system piped water is fed to water turbines which in turn operate AC electrical generators. Thus in a typical scheme a dam is built across a river at a suitable point that will impound water over a large area. It is interesting to note that Hydro-electricity supplies nearly 20 per cent of the world's electricity.

There are fundamentally two basic types of dam. The first type being the gravity dam which relies upon the weight of its own material to resist the forces imposed upon it. Then second type is the arch dam which employs an arch shape to take the forces in a horizontal direction into the sides of the river valley.

One of the world's most well-known arch dams is the Hoover Dam (known as the Boulder Dam 1933-47) in Boulder Canyon, Nevada, USA. It is the highest concrete dam in the USA standing at 221 metres (726 ft) above the Colorado River. The power generating station at the Hoover Dam was completed in 1936 and has a hydro-electric power capacity of 1,300 megawatts.

The Grand Coulee Dam across the Columbia River, USA is one of the largest concrete structures in the world and is a hydro-electric dam that was completed in 1941. The Grand Coulee Dam is the largest producer of hydro-electricity in the USA, producing 6,800 megawatts. The Grand Coulee hydro-electric Dam is part of the Columbia River Hydro System, which is a series of 14 dams harnessing the energy of the Columbia

River, with the whole system generating 35,000 megawatts of electricity. It is claimed the dam is expected to last until the next ice age in 10,000 years; the hydro-electric facilities for a mere 500 years.

The world's most powerful dam, before China entered the arena, was the Itaipu Dam, Brazil and Paraguay generating 12,700 megawatts of electricity, but its crown has now been taken away by the Three Gorges Dam spanning the Yangtze River, at Sandouping, Yichang, Hubei province, China – the dam is as wide and twice as tall as the Golden Gate Bridge in San Fransisco - and is the world's largest power station in terms of installed capacity (22,500 MW).

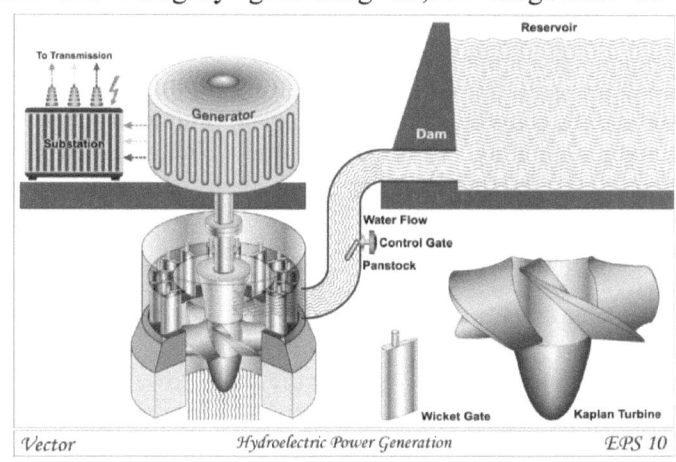

The dam is the largest operating hydroelectric facility in terms of annual energy generation, generating 83.7 TWh in 2013 and 98.8 TWh in 2014, while the annual energy generation of the Itaipú Dam in Brazil and Paraguay was 98.6 TWh in 2013. It must be said that the construction of this dam has been quite a controversial undertaking as it has displaced a claimed 1.9 million people, not to mention the loss of many valuable archaeological and cultural sites, as well as the effects on the local environment. The proponents of the project counter this criticism by pointing to the economic benefits from the subsequent flood control and hydro-electric power.

The largest Hydro-power station in England and Wales is the E.ON Power (formerly Powergen) Cwm Rheidol hydro-electric scheme located near Capel Bangor, nine miles from Aberystwyth in mid-Wales. The construction work began in 1957 and the scheme commenced operation in 1961, being officially opened in July 1964. The total capacity of the scheme is 56 MW and it generates on average four hours a day; the scheme can reach full load within twelve minutes.

Cwm Rheidol, although the largest hydro scheme in England and Wales, is obviously just not in the same league as the world's larger hydro-electric schemes. This is simply because the UK does not have the necessary large rivers and required fall of water. Nevertheless, improvements in small turbines and generator technology will make 'micro' (under 100 kW) hydro schemes a much more attractive means of generating electricity.

These small schemes have a range of anything from a few hundred watts for small domestic schemes to a minimum of 25 kW for commercial schemes. Indeed, useful power may be harnessed from even a small stream: a small turbine on a hill stream could generate about 1 kW. It should be noted that micro-hydro schemes would use run-of-the-river technology rather than employing dams and storing large quantities of water. The run-of-river hydro schemes takes advantage of naturally occurring drops in a river or existing structures such as weirs.

These schemes do not interrupt the river flow, but divert part of it through a channel or pipe into a turbine, ensuring minimal impact on the environment.

The following information relating to the Garneth hydro-electric scheme has been obtained from the Powergen website and reproduced here by kind courtesy of E.ON, and I quote, "The Garneth hydro-electric scheme is a run-of-river hydro scheme situated at Pen-y-Bont, Dolwyddelan, in the Snowdonia National Park. The project uses a water turbine to exploit the energy potential of the Tyn-y-ddol River. The scheme was commissioned in 1992, has a maximum output of 580 kW and is owned by Garnedd Power Company Ltd, being a joint venture between Powergen (now E.ON) and the Weiss Group. Garnedd is unmanned and is monitored remotely from E.ON's Rheidol hydro-electric station. The annual generation (based on DETR figures) meets the needs of around 360 homes and Garnedd is estimated to save around 660 tonnes of carbon dioxide a year. The system briefly works as follows: Leaving an agreed flow in the river, water is diverted from a small head pond into a 600 mm cast-iron buried pipeline, approximately 1 kilometre in length. The static head of the scheme is 102 metres - a Turgo impulse turbine drives a synchronous generator. During construction some trees had to be felled to establish the pipeline route but this was incorporated into normal forestry thinning operations. Ground disturbed during the construction of the pipeline was grassed over and became an effective firebreak. The turbine house has been faced with local stone and roofed in Welsh slate to blend in with its surroundings." End of quote.

When one considers the number of rivers in the UK a considerable amount of sustainable, environmentally friendly electricity could be produced, replacing ineffective wind generators and supplementing the electricity generated from other sources.

Geothermal Energy

Globally there are five types of geothermal energy such as hot water, hot dry rocks, magma, compressed hot water aquifers and ground-source heat. I would imagine that the hot water kind is the one most people are aware of – whereby heat from hot, briny water deep below the Earth's crust is brought to the surface as steam or hot water, then used to generate electricity by virtue of a steam turbine before it is returned to the wells. The production of electricity from geothermal energy was first produced at Larderello, Italy in 1904; today the use of geothermal electricity has grown to approximately 7,000 megawatts in twenty-one countries around the world, with the USA contributing with nearly 3,000 megawatts. Not only can geothermal energy produce electricity but obviously has other uses such as heating buildings - the first town in the world to heat its buildings with geothermal energy was Boise, Idaho, USA, in 1892 using hot water from underground springs. Unfortunately for the UK we do not have hot springs on the same scale as in other parts of the world, but we can still exploit ground-source heat in our efforts for environmentally friendly sources of energy.

The temperature a few metres below the surface of the Earth in the UK, keeps at a fairly constant level of about 10 degrees Celsius throughout the year. This heat, by means of suitable underground pipes and geothermal heat pumps can be extracted for space heating in buildings, and in some cases, pre-heating domestic hot water. By utilising ground-sourced heat we can indirectly reduce our requirement for electrical energy and therefore contribute to the saving in atmospheric emissions.

Fuel Cells

It may come as a surprise to some readers but the concept of the fuel cell was invented by a Swansea man by the name of Sir William Robert Grove (1811-1896) way back in 1839; he became known as the 'Father of the Fuel Cell' although the term 'fuel cell' is attributed to Ludwig Mond and Charles Langer who attempted to build the first practical cell by using air and industrial coal gas. In the early 1960's NASA employed fuel cell based electrical power systems in their space craft rather than using solar power as they deemed it too bulky, nor did they consider nuclear reactors because of the high risk factor. Indeed, the Gemini and Apollo space capsules utilised fuel cells produced by General Electric.

During the 1960's through to the 1980's fuel cells were too expensive for general applications, but then in 1990 large investments were poured into fuel cell companies, and although it can be argued that it was the space industry that has brought fuel cell technology to its current level, it was this additional investment that was instrumental in making fuel cells viable for the private sector.

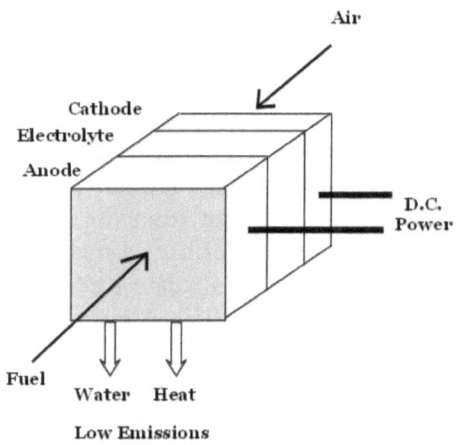

In this simple diagram of a fuel cell on the one side (anode) we have fuel in the form of hydrogen gas, and on the other side (cathode) is oxygen (in air). Sandwiched between the anode and cathode is a very thin, gas tight, electrically insulating but ion conducting, electrolyte layer. If the anode and cathode are connected to an electrical circuit then a D.C. Current will flow.

A fuel cell is a device that produces electrical energy. They operate by using the chemical properties of hydrogen and oxygen to create a useable electrical current, heat and water vapour. In some ways they can be considered to be a continuously charged or fuelled battery, although, unlike a battery, they do not store electrical energy, but continuously convert chemical energy to electrical energy.

Fuel cells are, for example, significantly more efficient than equivalent Internal Combustion Engines (ICEs) because they convert chemical energy directly to electrical energy rather than via an efficiency sapping mechanical intermediate phase. They operate at maximum efficiency at part load (where most ICE generators operate) and their efficiency is unaffected by size. In addition, the modular design allows the fuel cells to be stacked in such a way to match the specific output power needs without significant additional design work or capital requirements.

The simple nature of their design and operation makes them highly reliable. They operate quietly and can function on a variety of fuel types. If pure hydrogen is used as a fuel then the only outputs are electricity, heat and water vapour. In this way fuel cells are seen as being significantly more environmentally friendly than other hydrocarbon fuelled power sources.

The Chemistry of Operation of a Fuel Cell

On one side (the anode side) of the fuel cell is fuel in the form of hydrogen gas, whilst on the other side (cathode side) is oxygen (in air). Sandwiched between the anode and cathode is the very thin, gas tight, electrically insulating but ion conducting, electrolyte layer. An electrical circuit connects the anode and provides the mechanism to power electrical devices. The combination of the materials used to make the fuel cell components, the type of fuel used and the operating temperatures allow electricity to be generated via a chemical reaction rather than burning the fuel.

The reaction starts with the oxygen on the cathode side being ionised at the cathode and generating negatively charged oxygen ions that then flow through the cathode and across the electrolyte. At the anode side the oxygen ion combines with a positively charged hydrogen ion and releases an electron that then, because of the charge imbalance and the electron-impermeable electrolyte, flows around the electrical circuit to the cathode side generating direct current.

This high quality, direct current will continue to be produced as long as there is a supply of fuel and air to the fuel cell. Therefore fuel cells offer a better way of producing electricity and heat than traditional power generating products. Fuel cells are highly efficient, quiet and produce low emissions – making it possible to move to a more secure and sustainable energy world. Up until now the choice has been basically between complex, delicate systems that operate at extremely high temperatures or vulnerable low temperature systems that can only run on pure hydrogen.

A company called Ceres Power Ltd and based in Crawley has developed a revolutionary fuel cell – the world's first commercial metal-supported solid oxide fuel cell operating at an intermediate temperature. It starts producing energy extremely quickly and is designed to work with a range of fuels, including LPG, natural gas, methanol, hydrogen and vehicle fuels. Thus the combination of traits makes it ideally suited for mass market use - a fuel cell that offers durability, efficiency and fast response, and all at a lower cost. Ceres Power Ltd is a high growth product development company founded in 2001 to commercially exploit revolutionary fuel cell technology originally developed within Imperial College during the preceding 10 years. Ceres is a product development company specialising in small scale distributed power generation using intermediate temperature solid oxide fuel cells. A single fuel cell is the basic building block of a fuel cell power generator; Ceres is developing low cost and robust fuel cells that will be combined into stacks capable of generating between 1 kW and 25 kW.

Nuclear (Fission)

My reaction to nuclear (fission) power stations is simply, why lay on a dirty and muddy beach which charges a high entrance fee, when you can sunbathe on a beach with soft golden sands and which has a far cheaper entrance fee? I am also reminded of the quote attributed to John Glenn, American Astronaut, which offered the profound words, "As I hurtled through space, one thought kept crossing my mind - every part of this rocket was supplied by the lowest bidder."

Apart from astronomical building costs – nuclear power plants can cost up to 30 times more than conventional coal or gas-fired (CCGT) power stations - there is also the horrendous de-commissioning, waste storage and disposal costs – do not underestimate the magnitude of these costs as there is not a single nuclear power plant anywhere on the planet that has been completely and safely decommissioned yet. Indeed, it is a very sobering thought to realise that all the nuclear plants that have been shut in approximately the last 30 years are still going through the stages of storage and processing of spent fuel and materials.

It is vital to appreciate the problems with hazardous materials (what a legacy for future generations) and the potential for a nasty mishap (leakage of radio-active material at the very least). It should be fully recognised that nuclear power plants are designed, built and will be maintained by Homo sapien...a species that does not have an impressive record, being prone to arrogance, pride, greed, corruption, miscalculation, error, accident, acts of war and terrorism - the historical record stands testimony to our current energy problems, and for those who can remember when nuclear power stations were first developed, we were promised that electricity would become so cheap no one would even bother metering it – with hindsight we can appreciate how hollow and meaningless this promise was. History is full of false promises and claims such that, it cannot happen to us – impossible - we have got it right, you just don't understand the science, and so on, ad infinitum.

Howard Aiken, the computer pioneer, is credited with having stated that only six digital computers would be required to satisfy the computing needs of the whole of the United States. The designers of the Titanic thought they had got it right - but history has shown that apart from questionable design (always easy with hindsight) - they ignored and treated nature with contempt.

Regarding other usages of nuclear energy how many readers remember, or are even aware of the near thermonuclear disaster off the Andalucian coast of Spain when on 17th January, 1966, a US Air Force giant B-52G bomber from the Seymour Johnson Air Force Base, North Carolina, USA, a collided with a KC-135 tanker aircraft whilst attempting to refuel at 9,450 metres (31,000 ft) causing a massive fireball which engulfed both aircraft – killing all four men on the tanker and three of the bomber's crew – four other bomber crew managed to eject before plunging into the sea where they were thankfully rescued by Spanish fishermen.

Later the official investigation concluded that the B-52G overran in manoeuvres to hook up with the trailing fuel boom and rammed the tanker. This aerial accident was horrendous by itself, but what made it a thousand times more terrifying was that the B-52G bomber was carrying four 1.5 megaton Mark 28 thermonuclear hydrogen bombs – each 70 times more powerful than the atomic bomb that destroyed Hiroshima in Japan in 1945.

Imagine the holocaust had the bombs detonated causing thermonuclear explosions – it would surely have been truly horrendous – but luckily they didn't - with one bomb splashing into the Mediterranean sea, another drifted down on its parachute landing in a dried-up river bed, with the other two splitting open on impact with the ground scattering plutonium and covering the tomato fields near the village of Palomares, Spain with a fine and deadly radioactive dust.

Plutonium is a very toxic material and takes thousands of years to become safe, and that is why air monitors have been installed around Palomares since 1966 and crops and animals are regularly examined – with hundreds of villagers undergoing numerous physical examination - despite efforts by the USA to clear up the contamination thousands of cubic metres of contaminated soil remain in the area – it is now nearly half a century since this incident and you have to wonder what it will take to make the area safe again for not only agriculture but for human habitation?

If Human's cannot be trusted with weapons of mass destruction such as hydrogen bombs, which by their very nature, demand extremely high standards of care and attention, then what of nuclear power stations placed on a relatively small island such as the UK.

It is not rocket science to work out that the building of a nuclear (fission) power station is just not worth the risk, coupled with its horrendous expense, when there are other viable options - remember it takes about 10 years to complete a nuclear station, and costs up to 30 times more than a coal or gas-fired power station. We should all be very reluctant to even contemplate the nuclear option in the provision of electrical energy - in fact we should not consider nuclear power as an alternative to anything unless there is a 100 per cent assurance to its safe use – and this is fundamentally not possible. It is more than evident that all systems designed and built by Humans are open to error or failure as all Human effort is prone to error or failure - we are all fallible.

Apart from horrendous cost it is this fundamental Human weakness that puts me in fear of Nuclear (fission) Power Stations – and the more of them, obviously, the greater the risk.

The Three Mile Island disaster (1979) happened for its own *specific* reasons - such as the combination of mechanical and electrical failure as well as operator error - causing a pressurised water reactor to leak radioactive matter.

The Chernobyl disaster (1986) occurred because of its own *specific* reason - overheating causing an explosive leak of clouds of radioactive material, resulting in large scale evacuations from Chernobyl and Pripyat - the next significant Nuclear Power Station disaster will occur due to its own *specific* reasons...

Such as the 2011 disaster at the Fukushima nuclear power plant near Iwaki, Japan when three nuclear reactors went into meltdown as a result of a 9.0 magnitude earthquake causing a tsunami that engulfed the power station. Now who would have dreamt of siting a nuclear power station in an area prone to earthquakes - especially directly on a coastline that faces out to the Pacific Ocean – an ocean that is no stranger to tsunamis - no doubt an Alien from another planet might hazard a guess as to who might be silly enough...?

What does it take to learn the lesson when more than 1,600 people have died from health complications brought on by this particular nuclear disaster in 2011 – since then the company that owns the plant, Tokyo Electric Power (Tepco), has struggled to bring the plant under control, with hundreds of litres of radioactive water flowing into the Pacific ocean – with the clean-up effort set to cost more than £11bn - excluding the compensation still owed to thousands of families.

It would appear the lesson has not been learnt by the UK government with the proposed new nuclear power station, Hinkley Point C (3.2 GW capacity with two reactors), which is situated in Somerset on the Severn Estuary approximately five miles from Bridgwater, 15 miles from Minehead in the west and roughly six miles from Burnham-on-Sea.

It should be noted there is the existing nuclear power station at Hinkley Point, namely Hinkley Point B power station - this was the first Advanced Gas-cooled Reactor to generate electricity to the grid in the UK and consisted of two Advanced Gas-cooled Reactors supplying a total of 955 MW to the National Grid. Construction started in 1967 and generation started in 1976 – a total of 9 years from start of construction to actual generation and the estimated decommissioning date is given for 2023.

Then during March 2009, Hinkley Point C was officially nominated as a potential site for a new nuclear power station by the Government. Unbelievably to be constructed on the coast, and low lying coast at that, in the Severn estuary – has the Japanese Fukushima nuclear plant disaster fallen on deaf ears – and is government that totally ignorant of our history?

On 30 January 1607, the Severn Estuary was devastated by a great flood which was particularly severe on the Welsh side, extending from Laugharne in Carmarthenshire to above Chepstow in Monmouthshire. The coasts of Devon and in particular, the Somerset Levels as far inland as Glastonbury Tor, 23 kilometres (14 miles) from the coast, were also affected. The sea wall at Burnham-on-Sea collapsed, and the water flowed over the low lying levels and moors – so what hope, if this happened again, for any low lying coastal nuclear power stations at Hinkley Point – Armageddon is the word that quickly springs to mind - the inundation of a nuclear power station (possibly two) and its consequent meltdown would make the 1607 disaster look like a walk in the park.

The 1607 flood affected thirty villages in Somerset, including Brean which was 'swallowed up' and where seven out of the nine houses were destroyed with 26 of the inhabitants dying – across the area an estimated 2,000 or more people drowned, with houses and villages swept away - 200 square miles (51,800 ha) of farmland inundated and livestock destroyed. The reasons for the great flood of 1607 are attributed to either a tsunami - due to massive undersea boulders being displaced resulting in rock erosion and causing high wave velocities throughout the Severn Estuary, or a storm surge - similar to the 1953 North Sea Flood where high tides and a storm surge resulted in floods in East Anglia, Canvey Island, the Netherlands, Belgium, and Scotland.

So I guess it is erring on the side of stupidity to think we are all safe from a disaster at a nuclear power station – and an ignorance of history means we will bound to make the same mistakes of history - there will be another mishap - it could be due to Nature or Human or equipment failure, a combination of events, or maybe an act of terrorism - it is just a question of time - the terrifying thought is how widespread and disastrous will it be? Some will argue that more deaths have occurred within the conventional power generating industry than at nuclear plants - but the power industry has been going since the 1880's, whilst the nuclear industry has only been running since the opening of Calder Hall by Queen Elizabeth 2 in 1956 (world's first nuclear power station).

It is somewhat pointless in comparing the accident statistics of the two industries as it will only take one major disaster at a nuclear plant in the UK, with the potential of killing and injuring (radiation) hundreds of thousands, or maybe millions of people and contaminating acres and acres of land to make a nonsense of any claimed safety statistics - do we really need to take this risk on our tiny and crowded island, when there are so many other options open to us? In a sense the 'real' question is not whether or not we need nuclear power, but can we afford a mishap on our tiny and crowded island - large land masses such as continental America or Russia may be willing to take the risk - they have the space - and have indeed just managed to escape an Armageddon type scenario so far – and it is fingers crossed for all other existing nuclear plants, especially those across the English Channel in France that a major incidence does not take place, after all, France is not that far away being too close for comfort.

At the end of the day common sense dictates that we are better placed taking our chances with emissions from a mix of combined cycle gas turbine stations, and a number of oil and coal-fired power stations with scrubbers than with the nuclear option...remember, as pointed out above, it will only take one major disaster to prove the point. But the supporters of nuclear energy will say, "Look we have come a long way since the first nuclear plant at Calder Hall, the technology has vastly improved, we have very much improved safety and we have learnt from the American and Russian incidents et cetera." To all this I would say 'poppy cock', and point to Japan - history clearly shows that the only lesson we learn from history, is that we never learn.

They said the Titanic was unsinkable, but human failing ruled the day when the captain said, "Keep me informed of icebergs, but it is full steam ahead." Could the Tacoma road bridge collapse in the United States have been prevented, or the collapse of the Tay Bridge in Scotland with its unfortunate passenger train - what of the mishaps attributable to nuclear submarines – then what of the space shuttle disasters and the loss of two Mars Polar Landers' during 1999 at a cost of 250 million dollars. Indeed, one of the Mars Polar Landers' failed as a result of - wait for it – a number of people working in Metric whilst others were using Imperial measurements, with the consequence of lost navigation in the space craft and 80 million dollars down the drain – they say life is stranger than fiction – and events of this nature certainly prove it.

It is also interesting to note when assessing risks and taking air travel as an example, you will find that air accidents are often caused by a chain of events, each relatively harmless in itself, but catastrophic when linked together. Reasons for plane crashes as sourced from the Aircraft Crashes Record Office are as follows, Human error: 68 per cent, Technical failure: 20 per cent, Weather: 6 per cent, Sabotage: 3 per cent and Freak causes: 3 per cent.

The miraculous escape of flight BA038, a Boeing 777, carrying 150 passengers and crew, at Heathrow airport on Thursday 17th, January, 2008 should concentrate minds to the skills and also, unfortunately, the fallibility of Homo sapien and their activities. Experts claimed that the chances of a double engine failure occurring on a modern aircraft are 'a million to one' – and although a catastrophe was avoided by the skill of the crew, it should be fully recognised that both engines did indeed fail simultaneously and at a very critical point in the flight! It can be argued that it was lucky the plane failed just at that particular moment in the flight (slightly earlier, then the aircraft may not have made it to the airport, coming down on a very congested urban area), the weather was good and the aircraft was not fully loaded; the 209 ft-long plane can carry between 305 and 440 passengers at a cruising speed of 615 mph with a range of up to 8,300 miles.

It was reported in the media at the time that the Boeing 777 was launched in June, 1995 and is considered an extremely reliable aircraft with an almost impeccable safety record; the plane is powered by two Rolls Royce engines and should still fly if one fails – its landing gear is the largest of any commercial aircraft.

I quote the above aircraft incident as it is a timely reminder to the advice of 'experts' and what could have been a major disaster killing hundreds of people – an incident at a nuclear power station has the potential to infect (radiation) and/or kill thousands, if not millions.

It is also instrumental to recall the tale that Robert Gates, the American defence secretary can offer about Zbigniew Brzezinski (when he was President Carter's national security adviser) being woken up with the news that 200 Soviet missiles were on their way to America. Brzezinski demanded confirmation before alerting the president. Two minutes later, he was informed that the radar now showed 2,000 missiles. Just before he woke Carter to tell him he had perhaps two minutes to launch a retaliatory strike, he was telephoned again to be told there had been an error and that someone had put an exercise tape into the computers by mistake. (With acknowledgements to the book, 'Arsenals of Folly: The Making of the Nuclear Arms Race', by Richard Rhodes).

Therefore you will no doubt understand my fears and not having too much faith in Human reassurances and guarantees. With hindsight, most, if not all past disasters could have been avoided, so let us not fall into the trap and arrogantly think that we now know it all – even with the best of training and the latest technology we do not (nor ever will) have all the answers.

Defenders of nuclear plants will point to a World Health Organisation (WHO) survey of the reactor incident at Chernobyl, Ukraine in 1986, which claims killed only 75 people, and most of them either operating the plant at the time or rescue workers at the scene - although it is recognised the zone around the plant, evacuated 20 years ago, has now become a thriving nature reserve. Now whether the WHO figure is correct or not I cannot say, although I suspect a lot more suffered, if not fatally, but from numerous associated radiation sicknesses.

Nevertheless, there was an incident and this is the important point - next time it may be far worse.

It is wise to keep in mind the words, "Never, say never." A similar claim may be made for the 1979 accident at Three Mile Island, as the number of deaths resulting seems to have been zero - or were they? But what of associated radiation illnesses - were there none? Nevertheless, again, an incident did happen! The following examples of further different incidents do not exactly fill one with confidence:

- ❖ It is unnerving to realise that in 1965 a US Navy Skyhawk jet bomber fell off the deck of a ship and Sunk in 4,900 metres (16,000 ft) of water, 130 kilometres (80 miles) off the coast of Japan - the aircraft took a one-megaton hydrogen bomb along with it to Davy Jones's Locker.

- ❖ At a uranium reprocessing plant at Tomsk, Siberia, Russia during April, 1993, a tank exploded sending a cloud of radioactive particles into the atmosphere.

- ❖ How many readers will recall that in 1957 at the Windscale (now renamed Sellafield), Cumbria, England, fire destroyed the core of a reactor, releasing large quantities of radioactive fumes into the atmosphere.

- ❖ With all the reassurances of the nuclear industry it is shocking that 380 million litres (100 million gallons) of radioactive water containing uranium leaked from a pond into the river Rio Purco, at Church Rock, New Mexico, USA in July 1979. The consequence of this was to cause the water to become over 6,500 times as radioactive as safety standards allow for drinking water.

- ❖ Scotland's Dounreay nuclear power station is expected to take until 2036 to dismantle at an estimated cost of 3 billion pounds. It is said to be a complicated job and how many people are aware of its poisonous past - during the 1960's and 1970's radioactive material created during the fuel processing found its way into the sea! Proponents of nuclear power will argue that was a long time ago and things have changed dramatically since then……….well really, then read on…

- ❖ During an incident in 2004 thousands of gallons of radioactive material leaked unnoticed from Sellafield's Thorp reprocessing plant – it was said to be the most serious leak anywhere in the world in 2004 – the problem was blamed on a faulty pipe which had suffered metal fatigue.

❖ Sunday Times, Oct 21, 2007 re "USA hits panic button as air force 'loses' nuclear missiles."

As mentioned above, the proponents of nuclear power will no doubt claim, we have come a long way since Calder Hall 1956 and Windscale 1957, we have superior technology and have learnt a lot of lessons - safety is paramount - to even think of any mishaps these days is being somewhat paranoid - so I guess I should not have mentioned the 2004 incident at Sellafield and the loss of a nuclear missile in 2007.

It could be argued that a breach in a large dam could easily kill tens of thousands and destroy hundreds of properties downstream, with even far greater numbers in the case of the giant Three Gorges project in central China - research from The Paul Scherrer Institute, a Swiss government physics research laboratory, indicates that nuclear power has been responsible for a tiny fraction of fatalities, a fortieth, of that for renewable hydroelectric power. But what are these arguments trying to say? As mentioned earlier, it will only take ONE major nuclear power station disaster to make a mockery of all these statistics - is this a chance you want to take on behalf of yourself and your children?

Supporters of nuclear energy will claim a significant reduction in atmospheric emissions if the UK changed to nuclear power – a claim similar to that of the supporters of wind generated electricity. But again, do these people not also realise that China is now the world's second largest producer of industrial energy and has a seven-year plan to build over 500 coal-fired power stations - think of all those potential emissions - the UK is responsible for less than 2 per cent of the world's emissions - and there is India, Mexico, Brazil - not to forget the USA. Even if you cannot agree with all or some of the above observations and consider the nuclear option is safe and worthy of consideration, you will still have to take on board the following:

The prohibitive cost of nuclear energy - the cost is in the billions of pounds and I doubt if there is anybody who actually knows the true total cost? Indeed, Guy Dauncey with Patrick Mazza in their book titled, "Stormy Weather", state that in the US, the nuclear industry has already received $145 billion in subsidies, compared to $5 billion for solar and wind energy. In the United Kingdom the Nuclear Decommissioning Authority (NDA) estimation for the decommissioning of the country's nuclear power sites would be of the order of £70 billion pounds! How many conventional power stations could be built for this amount of money, and how much environmentally friendly energy research and development would this had funded?

It really does make you think – what if this money could have been spent on developing advanced LED lighting and *energy saving devices*– a very worthwhile goal as it would lead to a quantum leap in electrical energy saving – think of that in environmental terms. To reiterate:

- Nuclear power stations are indeed highly expensive to build costing in the region of about £2 billion for a 1200 megawatt reactor, whilst the cost of a similar capacity gas-fired (CCGT) plant would be in the order of £400 million.

- The long-time scales involved in decommissioning old nuclear plants: It is estimated that Sellafield in Cumbria will take up to 75 years to clean up.

- The problem of what to do with nuclear waste! Nuclear waste stays highly radioactive for many thousands of years: a very real problem and future generations will not thank us for this.

- The possible use of plutonium for nuclear weapons.

- Acts of terrorism: it would be very foolish and irresponsible to consider a nuclear power station 100 per cent absolutely safe from acts of terrorism.

Finally, during May 2006 it was reported in the media that there have been 57 safety incidents around Britain since 1997 - the incidents ranging from equipment failure, radiation leaks to a fire and ground water contamination.

There are other safer, realistic options...so why do we have to go nuclear? (Pun intended).

Shale Gas (Fracking)

This chapter would not be complete without mentioning fracking and there has been a lot of misleading information written about the subject ranging from the poisoning of drinking water to the collapse of buildings due to earthquakes. It is amazing how many people will 'shout' against something having little or no understanding of the issue they are protesting about – take the case of contaminating groundwater with methane, fracking fluid, chemicals, and dissolved contaminants in flow water due to the activities of shale gas operations – do these people not realise that ground water is near the surface whilst shale gas lies some 2 kilometres to 5 kilometres below the surface – so unless the integrity of the drilling pipe (casing) fails, how can ground water become contaminated?

To put this into some kind of perspective, how many know the odds are greater for a person in London to be blown up as a result of a leaking gas pipe whilst walking along a pavement, than by accidently drinking water that has been contaminated by fracking. Services such as telephone and electricity cables, water and gas pipes are buried under the pavement at depths better measured in inches than feet, and a lot of these services are now many years old and are failing - how many know that the Health and Safety Executive is aware of a number of incidents in the London area in recent years involving exploding utility equipment, including foot-way box covers and electrical boxes. The majority of these incidents have not resulted in very serious or fatal injury so far, and have been attributed to electrical faults and/or gas leaks.

So what exactly is shale gas? According to the Geological Society of London (with acknowledgements) shale gas is extracted from fine-grained sedimentary rocks (called shale) formed over millions of years as the result of compaction of fine particles of mud – organic matter can be trapped in the layers as they are compacted, and are slowly converted through heat and pressure into hydrocarbons such as natural gas. The main component of shale gas, and the primary reason for its extraction, is methane, which makes up between 70 and 90 per cent of shale gas together with smaller amounts of other light hydrocarbons, carbon dioxide, oxygen, nitrogen, hydrogen sulphide, radon and rare gases. There are large sedimentary basins in the UK which contain significant shale sections.

Exploration for shale gas in the UK is still at an early stage, so there is currently no clear consensus about how much shale gas is under the ground and the prospects for extracting it economically. Nevertheless, most geologists agree that there are reasonably significant onshore resources, most likely in the Lower Carboniferous around the Pennines, in Jurassic layers in the Weald and Wessex, in the Upper Cambrian in the Midlands, and possibly in the Lower Palaeozoic black slate of Wales and South West England. Whether these resources are exploited will depend on economic, environmental, social, and regulatory constraints. There are large resources worldwide.

The Geological Society of London point out that there are risks and challenges associated with the extraction of any mineral resources, including shale gas. It is important that such activity is appropriately regulated, and risks identified and managed. Three areas of potential risk which have given rise to particular concern among policy-makers and the public are: groundwater contamination; water sourcing and disposal; and induced seismicity.

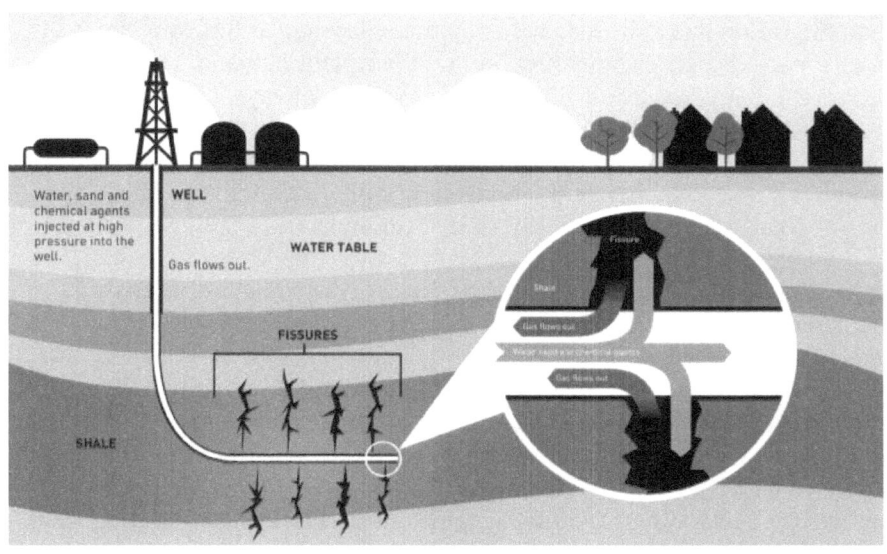

In the UK, groundwater provides 35 per cent of drinking water. Groundwater is also important to support surface water flow and regulate the health of ecosystems. Concerns have been raised (as mentioned above) about the possible contamination of groundwater by methane, fracking fluid chemicals, and dissolved contaminants in flowback water, as a result of shale gas operations. In the UK, most aquifers used for drinking water lie within the first 300 metres below the surface, while fracking operations would take place at a depth of more than two kilometres. Assuming wells are properly constructed, contamination of ground water through migration of methane and fracking fluids from shale formations to shallow aquifers through stimulated fractures could only take place if the fractures are able to propagate vertically through the intervening layers of rock.

Recent analysis of fracking operations in the USA, combined with data obtained from natural fracturing of rocks, indicates that the probability of a stimulated fracture exceeding a height of 350 meters is around 1 per cent. The analysis suggests that if a separation distance of at least 600 metres is maintained between aquifers and fracture zones, the risk of a fracture propagating to the aquifer and causing contamination is extremely low. Confidence in this result would be increased by conducting similar analyses for UK shale formations.

There are recorded instances of methane in groundwater in the USA in areas where shale gas operations have taken place. A more likely cause than migration through fractures is methane leakage at the well site itself, due to poor design or construction, or subsequent damage, (Historically, onshore US hydrocarbon operations have not always been effectively regulated, and in some areas there is a lack of records relating to well design and construction). Methane can also occur naturally in shallow groundwater. Geochemical analysis can distinguish this from thermogenic methane from deep shale formations.

Induced seismicity is the release of energy stored in the Earth's crust triggered by human activity, and is known to be caused by activities such as mining, deep quarrying, geothermal energy production and underground fluid disposal. During 2011, two seismic events of magnitude 2.3 and 1.5 took place in Lancashire, close to a fracking test site operated by Cuadrilla. Operations were suspended, and subsequent studies have suggested that hydraulic fracturing is likely to have been the cause, by reactivating an existing fault. The crust in most of the UK is relatively weak, and unable to store sufficient energy for large seismic events.

This means that the largest natural earthquake we can expect is likely to be no greater than magnitude 6. However, based on our understanding of the mechanical strength of shale and case studies of fracking operations in the USA, it is extremely unlikely that seismic events induced by fracking will ever reach a magnitude greater than 3. These are likely to be detectable by few people and are highly unlikely to cause any structural damage at the surface. To minimise the risk of seismic events even at this level, operators should avoid drilling through or near faults, and microseismicity should be monitored in real time before, during and after fracking, with effective management systems in place to respond to the results, including monitoring possible damage to well integrity.

The reader should recognise that there are in the region of 200-300 earthquakes in the UK every year, although the majority are so small that no one is aware of them with only about 10 per cent being felt by people each year. Nevertheless, a number of UK earthquakes have caused damage with the largest known earthquake (with a magnitude of 6.1) occurring near the Dogger Bank in 1931 – although it occurred 60 miles offshore it still caused minor damage to buildings on the East Coast of England. During 1884 the most damaging earthquake took place in Colchester where over 1000 buildings needed repairs with chimneys collapsing and walls cracking. Recently, during February 2015, an earthquake of magnitude 2.9 occurred in the Channel Islands which resulted in more than 100 reports from alarmed residents. The British Geological Survey said the epicentre was located about 16 miles south-west of St. Helier.

Finally, the author maintains that until we truly put an end to our dependence on fossil fuels and resolve the energy crisis facing us, then fracking has to be more than an acceptable proposition for the UK – it has the potential to transform the economy of the nation, whilst at the same time eroding our dependence on imported natural gas.

It is unbelievable that shale gas is taking so long to take off here in the UK whilst in the USA they are grabbing this 'bonanza' with both hands and transforming the American economy.

"I think there is a world market for maybe five computers."

Thomas Watson, Chairman of IBM, 1943

CHAPTER SIX

Future Power Generation and Consumption

To our everlasting shame we waste in the region of 30 per cent of our energy, so I guess that most self-respecting folk would agree that conservation should be high on the agenda and indeed, if at all possible, be made mandatory – or at least, with government remuneration as an incentive for recycling.

We live in a 'throw-away' and an extremely wasteful society to the extent that according to Love Food Hate Waste (http://england.lovefoodhatewaste.com) we throw away 7 million tonnes of food and drink from our homes every year in the UK, and more than half of this is food and drink we could have consumed. Almost 50 per cent of the total amount of food thrown away in the UK comes from our homes.

Love Food Hate Waste was launched in 2007 and raises awareness of the need to reduce food waste and helps us all take action to tackle it. It shows that by doing some easy practical everyday things in the home we can all waste less food, which will ultimately benefit our purses and the environment too. Since then avoidable food and drink waste (the good stuff that could once have been eaten) has reduced by a massive 21 per cent saving consumers £3.3 billion a year and councils around £85 million in 2012 alone.

Love Food Hate Waste is part of WRAP (www.wrap.org.uk), which is a registered charity (no. 1159512) and a Company limited by guarantee in England & Wales (no. 4125764) - WRAP works to help businesses and individuals reap the benefits of reducing waste, develop sustainable products and use resources in an efficient way.

The British Astronomical Association's Campaign for Dark Skies estimates that around 30 per cent of street lighting is shining up into the night sky, wasting £110 million each year, while total light wasted from homes, security lighting and floodlights costs over £1 billion per annum – sadly less than 10 per cent of the UK population can see the beauty of a natural night sky full of stars, let alone the majesty of the Milky Way – photographs taken from the International Space Station moving across the night side of the Earth truly show how light pollution has spread across the face of the planet.

Using energy more efficiently must be at the front line of efforts in caring for and truly looking after our only home, that of mother Earth. For example, research by Oxford University Environmental Change Institute has demonstrated how we can cut carbon dioxide emissions from our homes by 60 per cent through better building standards and improving the energy efficiency of existing houses, which are notoriously wasteful. At the time of writing only 18 per cent of UK homes are fully insulated, and yet our homes use 30 per cent of the UK's energy.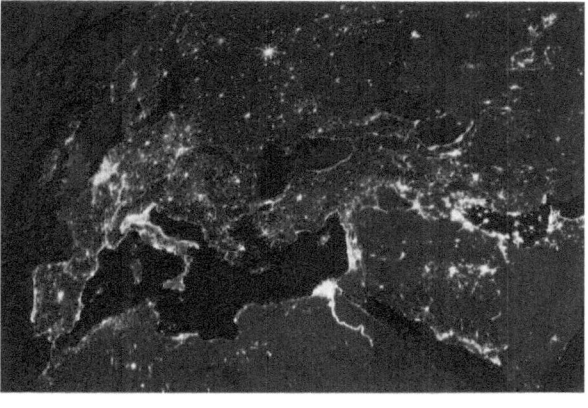
We can, and must, use fossil fuels much more efficiently such as in combined heat and power plants known as Co-generation (see later in this chapter).

In the very short term, existing coal-fired power stations could be refitted with higher efficiency boilers, and given the other options open to generating electrical power, is it any wonder the big environmental groups are united in their opposition to new nuclear (fission) power stations - this contentious, hazardous and expensive source of electricity would take more than 10 years (from planning, to building and commissioning new plants) to make a contribution - a crucial 10 years when our energy requirements need addressing now!

Regarding households are we that blind to the lighting and heating issues of the very homes we live in - it is a sad indictment that the most widely used electrical device to illuminate our homes up and into the 21st

Century has been the incandescent bulb – a sobering fact to realise this bulb was invented nearly one hundred and fifty years ago in the 1880's - it is only about 3 per cent efficient and it is not before time that it is truly being laid to rest by the mandatory use (2012) of the lower energy bulbs (LEB's).

The end of the Incandescent Light Bulb

The incandescent bulb was used as the normal means of illuminating the home during the 20th Century and it was recognised that lighting consumed 20-25 per cent of all electricity, thus it follows that a substantial reduction in the consumption of electrical energy could be achieved if an improvement in lighting efficiency could be found. Thus it was so, and it was with this step improvement a number of years ago I made the decision to change to low energy bulbs known as Compact Fluorescent Lamps (CFLs).

Although saving in power consumption the CFL low energy bulb has a number of contentious issues such as its relatively high cost and the problems in disposal when the CFL bulb had come to the end of its useful life. Unlike incandescent bulbs, the low energy fluorescent bulb contains potentially harmful chemicals. When the tungsten filament bulb reaches the end of its life span, there are not any significant problems with its disposal - it can be treated as regular waste - the only potential harm can come from careless handling in breaking its glass bulb. But low energy bulbs such as CFLs contain potentially harmful substances such as mercury - these bulbs obviously need to be handled with caution and care, in line with health and safety policies, and disposed of in accordance with the local waste authority rules.

There are now three main categories of light bulbs available:

1. Compact fluorescent lamps (CFLs) are currently the most common.

2. Halogen bulbs although the cheapest, are the least durable and energy efficient.

3. Light emitting diode (LED) lights are the most durable and efficient, but currently the most expensive.

As the halogen bulb is not truly a low energy bulb being the least durable and energy efficient it will not be discussed, indeed the author can envisage this bulb being consigned to history before long and being replaced by LED lights – a 50 watt halogen bulb (GU10 fitting) can be easily replaced by a 5 watt LED bulb (GU10 fitting) that uses only a tenth of the power consumed by the halogen bulb.

Compact Fluorescent Light (CFL)

The Compact Fluorescent Light (CFL) is considered, by the author, as simply a stepping stone to the next stage of TRULY EFFICIENT ENERGY SAVING ILLUMINATION - that of the Light Emitting Diode (LED) – see later in this chapter. It is claimed that the CFL could lose 20 per cent of its light over its 8000-hour lifespan - a CFL bulb gives off light when an electrical current passes through a gas-filled tube. The gas glows with ultraviolet radiation which lights up a coating of white phosphor on the inside of the tube, but eventually, this coating loses some of its ability to light up and hence the dimming.

Why do we need to illuminate so brightly the many business zones, housing estates, shopping centres, car parks, miles and miles of roadways, causing irritation to astronomers with all the unnecessary light pollution of the night sky - there may be arguments pertaining to security, but why such a degree of brilliance coupled with bad designs that waste energy - surely better design and the use of more efficient illuminating devices should now be more commonplace.

As mentioned earlier, unless light pollution is tackled realistically then our children and grandchildren will not be aware of the beauty of the night sky, for apart from the brightest objects such as the Moon, and possibly Venus, Jupiter and Sirius all other celestial objects will be a mystery to them – never the joy of spotting a shooting star and making a wish, nor the excitement of that rare visitor a comet – in the past comets have put on a magnificent show - sightings of manmade objects such as satellites or the International Space Station will also be impossible.

Then what of education and knowledge of the constellations and the ability to navigate by the stars – knowledge of where to look for Ursa Major (Great Bear) leading to spotting Polaris (North Star) will let the observer know where North is, and once this is known all the other cardinal points of the compass will be known. The majesty of the Milky Way will be a lost wonder to most, except those who live in remote and unlit parts of the Kingdom.

It should be remembered that 20th Century lighting consumed 20-25 per cent of all electricity, and although employment of CFL technology is a step in the right direction, we should now be moving swiftly to a much more effective technology - that of the Light Emitting Diode (LED) technology - this will afford a new era in the design and use of lighting in our homes and offices giving a very real reduction in the use of electrical energy for illumination.

Compact Fluorescent Light (CFL) versus the Incandescent Light Bulb

The design and implementation of a variety of low power devices can have a very significant impact on power usage - to the extent there will be no need to increase electrical energy production. In fact, it can result in the closure of a number of power stations. Therefore the intelligent thrust in overall energy conservation should be in minimising our electrical energy usage, coupled with a re-examination of our electrical producing industry.

So it will be useful, before proceeding to light emitting diode illumination, to examine the savings that the current use of compact fluorescent lamps can achieve.

It has been noted that 20th Century lighting consumed a significant amount of electricity, thus it follows that a substantial reduction in the consumption of electrical energy can be achieved by the use of low energy lighting - a compact fluorescent lamp will use only 20 watts of power to give the same level of lighting (brightness) comparable to a tungsten filament bulb of 100 watts, thereby realising a claimed saving of about 80 per cent.

Compact fluorescent lamps are not very popular, and for good reasons. They take a time to reach full illumination and then give out a cold light, whilst some folk claim that they flicker. In my experience compact fluorescent lamps used in the home environment have not been prone to flickering, but the time to reach full illumination could be irritating. There is also the problem to the disposing of faulty bulbs and the consideration of the small amount of mercury used in each bulb – one bulb may not be a problem, but a large number are.

Although CFL bulbs are not the Golden Fleece of energy saving lighting it is instructive to consider a comparison of incandescent and CFL lighting in a small country such as Wales. Thus consideration was given to the savings achievable in a single household by comparing the different energy consumed from using six 100-watt incandescent light bulbs to that of six 20-watt CFL bulbs between the hours of 4.00 pm and 11.00 pm each day (7 hours a day) for the winter months of October to March inclusive. The calculations revealed (see Appendix 3) that 756 kWh of electrical energy was consumed when the house was lit with the incandescent bulbs - but only 151.2 kWh was consumed when employing CFL bulbs – a saving (rounding up) of 605 kWh.

If (for sake of simplification) it is assumed there are 1 million dwellings in Wales (actual figure 1,279,494, see Appendix 3) then a saving of 605 million kWh (605 GWh) is possible for the whole of Wales – a staggering amount of electrical power saved during the winter months, with the potential to possibly saving half of that amount again during the summer period of the year.

At the time of writing modern 20-watt low CFL energy bulbs giving the same brightness as a 100-watt filament bulb could be obtained for about £1.00 each. It is interesting to ponder that the Welsh Assembly would (or should) have extremely good bargaining powers in wishing to purchase 6 million energy saving CFL bulbs and should be able to negotiate with the manufacturer/supplier a price for each bulb of at least 50 pence. Thus, if for example, the Welsh Assembly did negotiate a price of 50 pence per bulb then the total figure for the purchase of 6 million CFL bulbs would obviously be reduced to £3 million. Therefore in pursuing a reduction in the carbon footprint for Wales it would surely make sense for the Welsh Government to provide each household in the Principality with free 6 CFL light bulbs as this would save at least the consumption of 605 GWh of electricity each year and a staggering 6 TWh of energy over ten years. A marvellous and meaningful gesture in reducing the carbon footprint for Wales – the industrialisation of acres of Welsh countryside by ineffective wind generators is judged 'Neanderthal' by comparison, and as we shall see later the energy savings will be far greater using LED bulbs.

Considering a 2 MW ineffective wind generator costs in the region of £1.5 M to £2 M, then for the cost of just two of these ineffective monstrosities the Welsh Assembly could achieve REAL and continuing annual electrical energy savings in Wales by making the generous offer of free bulbs – the calculations have taken into account only domestic consumption during the winter months and there will obviously be a greater saving over a twelve month period – what would be the saving if factories, offices and all street and highway lighting were factored in - it does make one wonder as to why local and central government are so misguided regarding the provision, use and conservation of energy.

If, dear reader, you deem the above savings somewhat fanciful then I would, respectfully, refer you to an excellent book titled 'Stormy Weather' by Guy Dauncy with Patrick Mazza which states that in Newcastle, Australia, the City Hall's Function Centre changed all their three hundred and eighty 100-watt incandescent bulbs to 18-watt compact fluorescent bulbs over a two year period - this act reduced the Centre's electrical energy consumption by an impressive 80 per cent, and recovered the cost within 2-3 years!

The ultimate economy would be in the use of the light-emitting diodes (LED's) for illumination - by comparison the CFL is quite *limited* when compared to the potential electrical energy savings and flexibility if LED's were employed for all types of lighting. Indeed, the LED can be considered to be a tiny light bulb without any of the disadvantages of the filament bulb – not being subject to wasteful heat and not having a filament that will burn out - they are illuminated purely by the movement of electrons in a semiconductor material – and they have a very long life span. To reiterate, the use of CFL low energy bulbs should be regarded as a short term option – not forgetting the problems in disposal when the CFL bulb has come to the end of its useful life.

The short term option in the use of CFL bulbs would allow for the full development of LED technology, culminating in huge savings in electrical energy usage – although I would suggest that this development might be frowned upon by the electricity companies - after all their business is to sell electrical power to their customers and keeping their shareholders happy. A big turn down in power requirement would not be welcomed from a profit point of view – as they say, turkeys do not vote for Christmas.

LED Illumination

Light Emitting Diodes (LEDs) are a marvellous technology that has the potential to transform illumination beyond current thinking and imagination - make no mistake, the future is LED. The way we illuminate our homes will truly see a renaissance such that walls will light up on our approach and dim as we pass, whole ceilings and walls will be illuminated, hand-held remote controls will change the colour of LED lighting - I have such spotlights  working now in the garden and they are extremely economical to run and very pleasing each with remote control to change colours – and not expensive to buy over the Internet at less than £15 each. Such is the technology there are now wireless light bulbs on the market that can also play music – these bulbs can be switched via a mobile telephone so offer total flexibility, that when on holiday, you can activate the bulb and give the impression your home is occupied thus deterring burglars, and possibly reducing home insurance. These bulbs are now available from major electrical retailers - an example being the MiPow Playbulb – although prices are high at around £50 and it may take a number of years to recoup the initial outlay, it is claimed households can save a significant amount of money per annum on electricity bills simply by using energy saving bulbs.

Regarding the option of LED lighting I have been assessing the Pharox 300 LED bulb at home and have been suitably impressed with its working. The bulb was produced by Lemnis Lighting (Technology Pioneer 2009 Award), The Netherlands. This marvellous bulb offered a 90 per cent energy saving and a lifetime of 25 years - being a very impressive device demanding only 6 watt of electrical power. To reiterate, the Pharox 300 LED bulb uses 90 per cent less power than its equivalent 60-watt incandescent bulb, coupled with having many more advantages than the current energy-saving light bulb (CFL) – such as being able to be used with a dimmer switch.

Lemnis Lighting puts sustainability at the centre of the design process - using this approach the company develops innovations that help consumers, companies and policy makers to replace traditional lighting with LED technology. The advantages of LED lighting offer more than energy savings alone: costs for maintenance and replacement are both reduced, while performance is enhanced.

Pharox LED is a Dutch innovation from (www.lemnislighting.com).
The following summarises the Pharox 300 benefits:

- ❖ Saves up to 90 per cent energy, compared to incandescent, up to 50 percent over CFL.

- ❖ Lifetime around 25 years, based on 4 hours of operation daily.

- ❖ Warm white light (+/- 3000K).

- ❖ Very energy efficient – 6 watt, replaces up to 60 watt incandescent.

- ❖ Dimmable.

- ❖ High light output (> 300 lumen).

- ❖ Fits most regular fixtures.

- ❖ Contains no mercury.

On their website Lemnis Lighting state and I quote, with acknowledgements,

"Lemnis Lighting is a forerunner in sustainable lighting with LED technology - Lemnis Lighting wants to enable the transition from current wasteful lighting solutions towards energy-efficient alternatives, without making concessions on customer needs. In order to have the biggest possible impact Lemnis wants to be a paragon of this transition towards energy-efficient solutions that are within reach for everyone on this world. In this way, Lemnis Lighting helps consumers, businesses and policy makers reduce their CO_2-footprint and save energy through new lighting solutions that employ super-energy efficient LEDs. You can see this philosophy at work in the lighting products that Lemnis develops. Through dedicated business units in the fields of home lighting, public lighting, greenhouse lighting and solar lighting, Lemnis markets high quality light that saves energy and lasts a generation. All of our products are best in their class, from street luminaires that improve sight to greenhouse lighting that makes plants grow bigger, from the best LED bulb in your house to the best solution for sustainable lighting in areas with no access to the power grid. Lighting is a necessity for education, commercial activity, recreation and safety.

Lemnis Lighting shows that there's a better approach to lighting, one that benefits your wallet and the planet alike. Lemnis LED technology offers:

- ❖ First-rate lighting for indoor and outdoor applications.

- ❖ Low energy usage, up to 85 per cent more efficient than conventional lighting.

- ❖ Long service life.

- ❖ Low maintenance cost.

- ❖ Big CO_2-reductions through efficient manufacturing and energy savings.

Lemnis Lighting B.V. is a Dutch company with limited liability, and is a subsidiary of Tendris Holding. Tendris initiates, develops and invests in companies that focus on sustainable, market oriented and environmentally friendly solutions." End of quote from the Lemnis Lighting website.

The only downside at the time of writing was the cost of £30.00 per bulb. But Lemnis Lighting now say the Pharox 300 is not available anymore, however they do have a nice broad range of replacements.

Costs are indeed coming down as recently I changed nine 50-watt halogen bulbs (GU10 fitting) in the kitchen for 5-watt LED (GU10 fitting) MiniSun bulbs which cost £5.75 each - these being obtainable as Cool White or Daylight illumination. This offers a saving of up to 405 watts when all the lights are on, and will return the capital outlay due to their efficiency and longevity.

MiniSun are a product of the LSE Retail Group Limited that sell lighting and light bulbs to lighting retailers and wholesalers, with the claimed attributes of the MiniSun as follows,

- ❖ Quality MiniSun Branded 5w SMD LED Light Bulb. Aluminium Construction and Protective Glass Lens, 120° Beam Angle.

- ❖ This Bulb Will Pay For Itself in Just Over 12 Months! (Assuming 1000 Hours Use at 15p per kWh).

- ❖ This Bulb Develops a Class Leading 520 Lumens. 35,000 Hour Bulb Life. Colour Rating: 6000K (Daylight/Cool White).

- ❖ Uses the Latest LED Chips (Twice as Bright as Many Competitors) - At Last a Viable Replacement for 50W Halogen Bulbs.

- ❖ Height: 58mm, Diameter: 50mm. One Year Guarantee against Faulty Manufacture - not Dimmable.

Having used these bulbs I would certainly recommend them.

To put the savings achievable by more economical lighting into context it is useful to compare the virtues of LEDs to limited wind technology, such that billions of pounds are being spent in the mad pursuit of large scale wind generation in the UK, which is not only an unforgivable waste of money, but a betrayal to all things decent as they not only industrialise and despoil the beautiful countryside in which they are built, but also reduce the value of properties in close proximity to these unwelcome monsters, and are a potential threat to human wellbeing. Future generations will surely look back with complete disbelief at our folly, and hold us all in contempt for our thoughtlessness - low energy lighting is one of the easiest and simplest ways to reduce not only the home power bill, but factory, office and street lighting bills as well.

Having compared the use of incandescent and CFL bulbs in Wales we will now assess the savings over the same winter months of October to March inclusive, using six 6-watt Pharox 300 bulbs instead of six 100-watt tungsten filament bulbs.

Calculations show (see Appendix 3) that in using six 6-watt Pharox 300 bulbs instead of six 100-watt incandescent bulbs only 45.36 Kwh would be consumed, instead of the 756 kWh for the incandescent bulbs - a saving of 710.6 kWh per household over the six month period.

Again rounding down the 1,279,494 council tax dwelling figure (for ease of calculation) to 1 million, then there is the potential to save at least (rounding up) 711 million kWh (711 GWh) for the winter six months – again a staggering amount of electrical power saved during the winter months, with the potential to possibly saving half of that amount again during the summer period of the year.

Therefore if we guestimate a saving of half again for the annual consumption then there would be a saving of 1066.5 GWh per annum, and this being a very conservative estimation.

Wales consumed 13,524 GWh during 2012 (see Appendix five) so just a change to LED lighting would save at least 8 per cent of Welsh electricity consumption – but what if the lighting for Factories, offices, street and highway lighting were converted to LED illumination - it is not difficult to realise the significant reduction there would be in consumption of electrical power.

Simply by considering domestic lighting alone we can see the immense saving in energy and cost that LED technology can offer – this saving could be greatly complemented if government and industry would only strive to develop and produce much lower energy demanding devices such as washing machines, ovens and electric kettles – it can be done if the will is there.

In my current home in Ceredigion (see next chapter) most of the lighting has been changed to that of the LED and taking the kitchen as an example, the nine 50-watt conventional GU10 bulbs have been replaced by nine 5-watt MiniSun LED GU10 bulbs, thus reducing a total of 450-watts to that of only 45-watts, a saving of 405-watts - an energy saving of 90 per cent by embracing new technology.

Liquid-Crystal Displays (LCD's)

At the time of writing, JVC was claiming to offer the world's thinnest LCD TV - the 42 inch and 46 inch models are 4cm thick at their widest point. The new slim LCD panel backlight unit is 40 per cent smaller in depth and bezel width, measuring just 20mm deep and 13mm wide. The unit weighs 12 kg and consumes just 145 watt. Apart from LED's and LCD's we have OLED's which are thin-film organic light-emitting diodes that are now being used in computer displays and television screens. Indeed, Sonys (www.sony.com) first OLED (Organic Light

Emitting Diode) Television, the XEL-1, has a 3 millimetre thin panel and offers unparalleled picture quality with amazing contrast and offers,

- ❖ Next-generation OLED HDTV technology.

- ❖ Full HD Colour spectrum and super deep black-levels.

- ❖ Incredibly Slim 3 mm Thin Panel.

- ❖ Lower power consumption than other TV technologies.

- ❖ 1,000,000:1 Contrast Ratio.

The cathode-ray tube (CRT) is effectively obsolete in the home as the video display device for computers and television sets have been replaced by liquid-crystal display (LCD) displays – many other items use LCD's such as digital clocks, watches, CD players and microwave ovens.

The cathode-ray tube required a positive anode voltage of 10 to 18 kV for monochrome pictures on screen sizes of 12 to 19 inches. For colour screens of 15 to 25 inches an anode voltage of 25 to 30 kV was required. This is in sharp contrast, for example, to the very low voltage level required for LCD's and to the forward-bias voltage of about 1.6 volts required to produce a forward current of about 20 mA in an LED.

It is beyond the scope of this book to fully examine the technology behind LED's, OLEDs and LCD's suffice to emphasise that the power consumption of these devices is extremely small in comparison to older technology light bulbs, computer monitors and television sets.

Therefore if we contemplate a house employing LED's for lighting, LCD's and OLEDs for television set screens and computer monitors, then the saving per household in electrical energy will be far greater than could be achievable with the old incandescent bulb and cathode-ray tube technology.

It is very sobering to realise that if the technology behind these electronic devices had experienced the same capital input as the nuclear industry, or indeed wind technology then, for example, the illumination of homes, offices and factories would have reached new dimensions long before now - actual walls, apart from ceilings, could be designed to emit light, and yes, even the floor - the level of illumination controlled by dimming devices, motion detectors, or light level detectors - the possibilities are endless and the energy consumption per building would, as we have seen, be extremely small.

So why do we need a new generation of nuclear (fission) power stations, the ridiculous ineffective wind farms and inappropriate solar parks – why are we still tolerating the excessive and wasteful usage of electrical energy, when there are so many other options - all it needs is vision, the will and bold leadership.

Indeed there would be no need to refrain from using the 'standby mode' of electronic equipment as compared to 'total consumption' it is an infinitesimal amount of energy - it is about getting things into perspective.

If manufacturers researched and developed lower energy demanding washing machines, clothes driers, dish washers, electric kettles and heating appliances, all this could offer a major saving in the demand for electrical power from every household – LG Electronics, a company in Slough, Berkshire has just introduced a Steam Direct Drive Washing Machine, which they claim, that apart from washing clothes and leaving them crease-free uses 35 per cent less water and 21 per cent less electricity than conventional models – what an excellent environmentally friendly and forward looking company – hopefully this will spur many other companies to follow their fine example.

Domestic Supply and Wiring

A modern domestic electricity supply typically consists of an underground or overhead service cable from the local power network, which is connected to the customer service head – being a sealed box containing the main supply fuse - this will typically have a value from 40–100 Amps. Separate live and neutral cables ('tails') are fed to an electricity meter, and often an earth conductor as well – in a lot of modern dwellings all are housed in a built-in box situated on the house external wall ensuring that the meter can be read externally without requiring access to the house - from the meter more tails proceed to the internal side of the house wall and into the consumer unit (also known as a fuse box) - on older houses the meter and consumer unit will be found inside the premises.

The consumer unit contains one or more main switches and an individual fuse or miniature circuit breaker (MCB) for each final circuit. Modern installations may use residual-current devices (RCDs) or residual current breakers with overcurrent protection (RCBOs). The RCDs are used for earth leakage protection, while RCBOs combine earth leakage protection with overcurrent protection.

UK domestic electrical circuits are normally described as either radial or ring. A radial circuit is one where power is transmitted from point to point by a single length of cable linking each point to the next. It starts at the consumer unit and simply terminates at the last connected device. It may branch at a connection point. Lighting circuits are normally wired in this way, but it may also be used for low power socket circuits.

In a ring circuit, a cable starts at the consumer unit and goes to each power socket in the same way as a radial circuit, but the last power socket is connected back to the supply so that the whole circuit forms a continuous ring. This means that there are two independent paths from the supply to every power socket. Ideally, the ring acts like two radial circuits proceeding in opposite directions around the ring, the dividing point between them dependent on the distribution of load in the ring. If the load is evenly split across the two directions, the current in each direction is half of the total, allowing the use of wire with half the current-carrying capacity. In practice, the load does not always split evenly, so thicker wire is used. It should be noted that large power consuming items such as an electrical stove or immersion heater will have their own direct wiring from the consumer unit.

Cables are most commonly a single outer sheath containing separately insulated live and neutral wires, and a non-insulated protective earth to which sleeving is added when exposed. Standard sizes have a conductor cross sectional area of 1, 1.5, 2.5, 4, 6 and 10 mm^2. Sizes of 1 or 1.5 mm^2 are typically used for 6 or 10 ampere lighting circuits and 2.5 mm^2 for socket circuits. The protective earth conductor in older cables was normally one standard size smaller than the main conductors but is now specified to be the same size.

The earthing conductor is uninsulated since it is not intended to have any voltage difference to surrounding earthed surfaces - additionally, if the insulation of a live or neutral wire becomes damaged, then the wire is more likely to earth itself on the bare earth conductor and in doing so either trip the RCD or burn the fuse out by drawing too much current.

With the advent of LED lighting technology it can be argued that a major re-think is needed in the household wiring requirement. Such that the household consumer unit should now only terminate heavy load requirements such as power sockets, electrical stoves, emersion heaters, electrical showers and any other heavy load device. All lighting circuits should be connected in a separate termination box that is connected to low voltage battery – yes, you have read that correctly – all the household lighting requirements will function quite easily and effectively from a low voltage battery, with the battery being recharged from the house AC mains supply (sockets). Indeed, if the house has solar panels on the roof then it can be reasoned that, apart from the initial capital cost, all lighting for the household will be cost free – remember the battery will be charged up during daylight hours (solar power) and satisfy LED needs for many, many hours.

The advantages of such a lighting system are as follows,

- ❖ Employment of lighter and much cheaper load handling cables for lighting circuits.

- ❖ Lighter and cheaper lighting sockets and switches.

- ❖ In the event of a mains power failure lighting is still available in the property.

- ❖ Security and safety is much enhanced.

- ❖ Use of LED technology ensures up to 90 per cent reduction in energy use.

- ❖ Smaller and cheaper consumer unit required for heavy duty domestic circuits.

- ❖ Minimum (leading to free) on-going lighting costs if house fitted with solar panels.

In the previous chapter we looked at various methods of producing environmentally friendly electrical energy other than nonsensical wind, and the arguments against nuclear power in the UK were also examined - but critics erroneously claim that we need large fossil fuelled and nuclear power stations connected to a large grid system to satisfy our current and increasing appetite for more and more electrical energy - but this is misguided - if we fully embrace modern technology such as LEDs for lighting and develop lower energy demanding white goods et cetera then our electrical energy demands will consequently fall.

Co-generation

One possibility away from the synchronised central power station system, the abomination of ineffective wind farms and inappropriate solar parks (the Sun does not shine at night or on cloudy days) is the consideration of co-generation. This is where a single source of fuel such as natural gas will drive an electrical generator and where the heat is used to produce steam or hot water. Industry has used co-generation for many years where both heat and electricity have been required; industry has used coal-fired boilers and steam turbines from which heat can be tapped.

Regarding domestic co-generation, a gas boiler which will provide heating and hot water, whilst at the same time providing electricity to power household appliances, has been designed by a ground-breaking technology British company called Flow Energy. This company has a research and development facility in Capenhurst, near Chester, a home energy business in Ipswich and a manufacturing facility in the Almond Valley, just outside Edinburgh – the company launched on the stock market in 2006, and the value of the company, at the time of writing, is just over £100m.

In conventional gas boilers the gas is burned to directly heat the water which is then run through pipes to provide household hot water, but also pumped around pipes and radiators to heat the home.

The Flow Energy boiler is called a microCHP boiler - CHP stands for Combined Heat and Power, and there are a few different kinds of microCHP technology but Flow Energy use the Organic Rankine Cycle. The Flow boiler comprises two parts – the heat-producing boiler itself and the electricity-generating power module, housed inside the boiler. Natural (mains) gas or LPG burns in the combustion chamber, just as it does in a standard boiler. But instead of directly heating water, it heats up a fluid in the power module. This fluid evaporates and the resulting vapour moves through a scroll expander, which spins and acts like a mini dynamo to generate electricity, which can be used in the home or exported to the electricity network. Once the vapour has moved through the scroll it condenses in a small heat exchanger to heat the water for the house central-heating system and the household hot water taps.

The launch version of the Flow boiler is suitable for 3-5 bedroom homes and it requires a separate hot water tank (it isn't a combi boiler) – although the company say they are launching a combi version later in 2015.

The Flow boiler is manufactured in a dedicated facility in Livingston, Scotland by Jabil Circuit - Jabil is a huge global company that has forged its reputation for the highest quality manufacturing by making products for some of the world's biggest brands. Jabil is a multi-billion pound company with a long history of manufacturing excellence.

The Main Features of the flow Energy boiler can be summed up as follows:

- ❖ Electricity-generating (microCHP) condensing gas boiler requiring a separate hot water tank.

- ❖ Generates around 2000 kWh of electricity a year.

- ❖ A-rated efficiency – 92 per cent.

- ❖ Wall mounted.

- ❖ 2 year boiler guarantee.

- ❖ Up to 10 year power module (electricity-generator) guarantee.

The revolutionary Flow Boiler costs £3675, including VAT - but Flow Energy say that if the customer signs up for their Finance or Freedom packages, they will receive £4800 over five years in reduced home energy bills - this means the Flow boiler will pay for itself – thus whilst other boilers may be cheaper, they don't provide a great return like the Flow boiler does - a standard installation costs £1800, including VAT, although every home is different and Flow Energy will provide a personal install quote.

Customers who may have difficulty in raising the capital for the boiler and an installation can take advantage of the unique finance package offered by the company which means the cost of a new boiler is avoided - once the Flow boiler is installed under the Flow Finance package, the company will deliver a reduction in the customer home energy bill of £4800 over five years - this allows the Flow boiler to pay for itself.

Thus the customer does not have to pay upfront for the Flow boiler as the company will organise a five year personal loan for its cost - the total cost of this loan over five years is £4529.89 (excluding installation) – remembering once the Flow boiler is installed, the company will reduce the home energy bill by £80 a month, and they will do this every month for the five years of the agreement - this means the home energy bill will be reduced by £960 a year, every year for five years – thus the total reduction in the home energy bill at the end of the five years will be £4800 – Flow Energy say the boiler will be installed and serviced by an experienced engineer from Mears Group - Mears have a 30 year history of caring for the heating systems of nearly a million homes in the UK, and Flow Energy have chosen Mears because of their reputation for excellence and shared values. Further details can be obtained from Flow Energy (www.flowenergy.uk.com).

No doubt the more astute reader will have recognised that if a household employed their own solar PV system and a co-generation system from Flow Energy, then summer and winter situations will be more than adequately covered - during the summer solar generation is at its maximum, whereas during the winter co-generation is at its maximum. Indeed, if the property had a 4 kW PV system installed in conjunction with the Flow Energy co-generation system, the house generation would be in excess of that of the average household consumption of about 3500-4500 kWh per annum, exporting the excess to the Grid.

Considering the country as a whole then larger co-generation systems can offer lots of possibilities as factories and offices could benefit from a single co-generation unit - a new housing estate could be supplied with just one large unit (thereby offering economy of scale) feeding each individual property with heat and electricity.

It can be argued that if the whole of the UK had a significant number of co-generation systems then the requirement for a National Grid System would greatly diminish doing away with many ugly pylons, cabling and sub-stations – thus if co-generation on a large scale were to be complimented with factory, office and home roof PV solar arrays, plus more use of tidal and hydro schemes complimented with a minimum number of cheap and efficient Combined Cycle Gas Turbine (CCGT) Power Stations, then there would be no need for coal, oil, or nuclear (fission) power stations, nor countryside industrialising wind farms and solar parks - in conjunction with these suggestions the ultimate aim should not only be to convert from AC to DC but to greatly reduce the voltage to a much safer and economical level.

Imagine, for example, a co-generation system situated in a suitable housing estate providing both electricity and heat - because there is no appreciable power loss in the cables feeding the houses the generator can be of the DC variety and operate at a safe voltage level - thus all the advantages and economies of a DC system would be realised, such as not having to provide transformers or rectifiers in electronic equipment and appliances.

As an *eco-evolutionary* step the best we can currently do is to employ CCGT power stations in a restricted and meaningful fashion - it will give a breathing space to fully develop such technology as fuel cells, possibly nuclear fusion power stations, and technologies that will maximise the intelligent use of solar radiation, geo-thermal and gravitational forces (tides) which will last for as long as the Sun shines and the Moon orbits the Earth.

The suggested co-generation scenario using a hybrid of localised natural gas turbines and solar panels, CCGT power stations, geo-thermal, tidal, and hydro schemes, would allow a return to a complete DC system as there would be no requirement to transmit HIGH VOLTAGE over long distances – to reiterate, the need for a large central synchronised AC infrastructure would be greatly diminished, as mentioned above, and as such would usher in an age of new innovative eco-friendly challenges.

We live on an island and most large cities are on, or close to the coast - near to potential power generation from the sea. For the immediate future shale gas should be exploited - CCGT power stations in conjunction with natural gas co-generation/roof solar arrays could initially cater for places such as large inland cities and towns – for the more rural areas, it should be remembered that gas can be stored in cylinders and tanks as they are now for households that run central heating, gas fires and cooking stoves on liquid natural gas (LNG).

If the country did convert to DC then initially there would inevitably be problems of compatibility, but this hiccup could easily be overcome in the changeover stages by employing INVERTERS in conjunction with co-generation systems et cetera. An inverter is a device for simply changing DC to AC and therefore by the employment of this device it will enable all AC items to work off the new DC system. Once the whole country has fully changed to DC the manufacturers can produce 'cheaper' and smaller electronic goods as the equipment, as already mentioned, will no longer need to house transformers or rectifiers to work on the DC system, which eventually means electronic good such as televisions, DVD players, satellite equipment, radios, home computers et cetera can be produced much cheaper – the environment and everybody benefits.

Smart Meters

Most folk must now be aware that Smart Meters are planned to be installed in every home in the UK by 2020 - Smart Meters are a replacement for existing meters, and can come in the form of gas or electricity meters and send electronic meter readings to your energy supplier automatically.

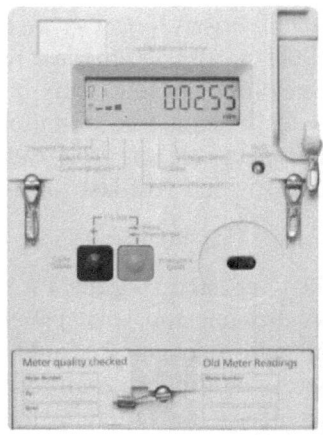

This book considers only electrical meters which work by communicating directly with the power company so the company will always have an accurate meter reading and there will be no need for the customer to take a meter reading. These meters can operate in a variety of different ways such as using an in-built SIM card (similar to the one in a mobile telephone) to send the readings, or can utilise long range radio. The claimed benefits for having a Smart Meter are simply that your meter can be read automatically (no need for a meter reader) and your bills will be up-to-date as there will be no more estimated bills and over, or under payments.

Although the meter by itself will not save the customer money it is claimed that such things as add-ons or computer software could – these additions enabling the customer to see at any one moment how much electrical power is being consumed, thus encouraging and enabling the customer to be more energy efficient. It is claimed that the meters could mean lower electricity bills as they will assist power companies to run more efficiently - surely this claim by any power company has to be tongue-in-cheek – lower electricity bills – when was the last time dear reader you experienced a meaningful reduction in your electricity bill from the profit making power companies?

The claim of reduced power bills by the power companies triggers the thought that if we all changed to low energy consuming electrical goods and appliances, whereby we all reduced our consumption by half, for example, then the electricity companies would have to double the cost of supplying electricity just to stand still from a profit point of view, otherwise they will obviously start to lose money, and the more we save the less they will earn – so where does the Smart Meter come in all this - the claim of the power companies that if they have an accurate feed-back of how much power is being consumed at any one time, then they will be able to supply the right amount of power at the right time is pretty near the truth - but Smart Meters have the potential to lead to the creation of much more *imaginative* new power tariffs, and I would suggest we all can guess what this means for the customer.

Surely, one of the most iniquitous charges to be found on UK electricity bills is the Standing Charge inasmuch it is not a *standard* charge - why do the electricity companies have different charges - at the time of writing Scottish Power are charging 20.55 pence per day for their Standing Charge, while EDF are charging 18.9 pence per day – recently I changed supplier from Scottish Power to EDF as the unit (kWh) charges were lower – but no engineers turned up to change the existing cabling – the external plant obviously stays the same, and I am still drawing my electricity from the same generators and network – so why this difference in price and the word obfuscation springs to mind.

If the Standing Charge were standardised or indeed abolished (being absorbed by the unit cost) then comparison between electricity companies on their unit (kWh) costs (the energy we use) would be made much more simpler – but perhaps the energy companies do not wish this, thus making it much more difficult for the customer to meaningfully compare costs – could I be forgiven for thinking the power companies are taking all of us for a soft touch?

Personally I cannot see what the customer gains from a Smart Meter as I have no difficulty in identifying my power usage with a conventional meter as it is not a demanding or mammoth task to read my electricity meter and notify the power company every couple of months – simply note the meter reading at the start of a billing period - again at the end of the billing period, then deducting the initial reading from the final reading to give the total power used, followed by a very simple calculation to find the cost.

As an example, if 1000 units were consumed over the billing period, then simply multiplying this figure by the power company cost per kWh (this information is on the power bill) will give the cost. Thus if the cost per unit happens to be 10 pence then the total cost would be, 1000 x 10 = £100 and if the customer consumed 4000 units over a year then the bill for the period would be £400, simples as they say.

Regarding the question of being aware of how much electricity is used on a daily basis and attempting to be energy efficient, it is not rocket science to realise that a 3 kW electric fire uses 30 times as much of power as a 100 watt incandescent light bulb, or to put it another way, if the 3 kW fire were to be run for 4 hours then a 100 watt bulb would run for 120 hours for the same power consumption.

All electrical devices have their voltage and wattage indicated and devices such as electrical irons, kettles, fires, washing machines, dish washers and spin driers are heavy users of power, whereas such things as radios, TVs, music centres and electrical bulbs use much less, and as we have seen LED lighting such as the Dutch Pharox 300 bulb consume very little power at 90 per cent efficiency.

I would also question whether the Smart Meter should have the ability to isolate a customer from the mains - if so, this will offer the power companies the ability to isolate a single, or any number of customers at any one time, and one wonders if this total control would be a good thing from the customer point of view.

With a conventional electricity meter and due to the very nature of the local distribution network (the sharing of local network power feeds) it is not easy to cut off single customers. Comparing this to the national telephone network, customers have a dedicated metallic path (a pair of wires) all the way back to their local telephone exchange thus enabling a customer, should the need arise, to be cut off at the telephone exchange - for example, a telephone user may have an electrical fault on their line and thus can be isolated at the exchange to protect telephone exchange equipment, whilst not affecting other telephone users. So although telephone landlines can be cut off remotely for individual customers, I am not too sure whether I would appreciate this control for power usage as I can possibly manage without the telephone, on say a cold winters day, but not so without electricity if someone decides, for whatever reason, to isolate the house from the mains supply – something for every reader to ponder on.

In the UK it is hoped Smart Meters will be in every home by 2020 – Smart Meters have been installed in many areas, in an effort to reduce electricity demand by making consumers more conscious of when and how they use energy – although I would argue that being in receipt of a hefty bill for excess energy usage has the ability to concentrate the mind much more so than any other means.

Summary and a Blue Print for the Future

➤ CCGT Power Stations

As regards UK electricity generation a sensible start on the *eco-evolutionary* ladder would be to implement a programme that called for *a specific* replacement of coal, oil and nuclear (fission) power stations as they age and come to the end of their useful life. This *specific* replacement being the more efficient gas fired (CCGT) power station, which is also quicker and much cheaper to build - indeed, it may prove advantageous to replace some of the other stations at an early stage.

➤ Co-generation

Government should consider subsidising the provision of co-generation in industry, offices and the home.

➤ Shale Gas

In conjunction with the above power station replacement programme and co-generation, the development of shale gas should be prioritised to meet the increasing demand of CCGT power stations and co-generation gas systems. This 'home source' of fuel which apart from ensuring security of supply will transform the whole of the UK economy as it has done across the Atlantic – the USA is benefiting hugely from its shale gas revolution going from a virtual running-out-of-gas scenario to one of abundance and export potential. Why is the UK dragging its feet to the potential of plentiful and cheap gas for our industries, enabling them to become much more globally competitive, the cheaper generation of electrical power and the heating of our homes – there is also the opportunity to power commercial transport and private vehicles – a golden gas bonanza truly awaits us all.

As mentioned earlier in this book there is a lot of misunderstanding and rubbish written about the fracking industry such as the danger of significant earthquakes, desecration of the countryside and contamination of ground water – negative propaganda which is grossly exaggerated – any ground tremors will hardly register on the Richter scale, disruption of the countryside will be minimal and undetected after installation and running of the site – contamination of ground water will not occur unless there is a failing in the bore hole casings, and with good engineering practices and regular inspections this is highly unlikely.

Fracking requires drilling bore holes to depths of greater than 7000 ft. down into the gas bearing shale, then drilling a six inch diameter horizontal well for several miles. Underground water supplies are to be found around 1000 ft. deep, so unless there is damage to the vertical bore casing at the aquifer level then there will no contamination of ground water. The nonsense spouted regarding earthquakes is put into perspective when one considers that in the millions of fracking operations in the USA there has not been one tremor big enough to cause damage. Remembering of course that the UK has numerous, natural small earth tremors each year which go unnoticed - fracking comes into this category.

Industrialisation of the countryside is another myth being pushed out by misinformed, engineering and technically challenged people, as once the initial drilling has finished the drilling rig will be removed, just leaving valves and pipes and a few possible enclosures no bigger than a site hut, such that hedging and trees will render the site invisible at ground level – with the site being possibly quieter than a library – the gentle hum of a site 'nodding donkey' tends to be drowned out by the sound of bird song. How many reader realise there are 50 year old fracking sites in the UK - these have not triggered any protests or marches and as such make the industrialisation claim utterly ridiculous – do the protesters of fracking not realise there has been fracking in Nottinghamshire since 1963 with one well subjected to fracking four times, yet there has been no environmental disaster – regarding environmental concerns the Royal Academy of Engineering and the chairman of the Environment Agency have declared fracking safe – what more assurance do the detractors need realising that since the USA commenced fracking, their carbon emissions have fallen to a 20 year low.

The abundance of shale gas will enable gas co-generation to come into its own with this cheap source of fuel and as more and more installations are completed the need for mains electricity will diminish reflecting in a run-down of large power stations - and the disappearance of long ugly pylon routes running across the countryside.

> Solar Energy

There are spaces on many industrial, office and domestic roof areas in the UK with the right orientation to exploit the energy of the Sun - although not 100 per cent effective due to our latitude, cloud cover and night time periods, roof-top PV and thermal solar panels, can contribute a significant amount of electrical and heat energy and as such have a place in the overall scheme of things. Indeed, with the government subsidy, domestic roof-top PV panels prove to be a good investment as detailed in the next chapter of this book.

> ➢ Severn Barrage

Government should implement large scale tidal generation schemes such as in the Bristol Channel, which has the second highest tidal range on the planet. Apart from tidal lagoons further down the channel, I would suggest building a barrage further up the channel for the generation of electricity – the barrage being also large enough to carry a rail crossing from Wales into England, thus making the Severn Railway Tunnel redundant with a consequent saving in pumping costs (the tunnel takes in water due to underground springs which were hit during the construction of the tunnel) and various other tunnel maintenance costs. Not only should the barrage be large enough for a railway crossing, but also that of a road crossing for the M4 constructed adjacent to, or on top, or under the rail crossing - this will also realise savings as one or both Severn Road bridges could be closed - savings in bridge maintenance plus offering road crossings that are not weather dependent as with the two existing bridges.

Imagine also an almond shaped island, with the narrow ends facing up and down channel, to be built half way across and intersecting the barrage, then a tourist attraction could be developed on this artificial island, which could consist of a hotel offering magnificent views up and down the Bristol Channel, tourist shops and cafes – a further great tourist attraction would be access to the 'Bristol Channel Barrage Generation Plant' similar to the Electric Mountain in North Wales. All this would encourage train excursions to run from Swansea, Cardiff, Newport, Hereford, Bristol, Birmingham and London et cetera – coach excursions from all parts of the UK, bringing in large crowds of *lucrative* tourists throughout the year – the scheme would have the potential of becoming one of the UK's main tourist attractions – possibly European as well – a true tourist magnet - it just needs the imagination, will and courage to grasp the nettle.

> Geothermal

There is an old saying that the answer lies in the soil and this has to be prophetic when it comes to geothermal heating systems. As mentioned earlier the temperature a few metres below the surface of the Earth keeps at a fairly constant level of about 10 degrees Celsius throughout the year. This heat, by means of suitable underground pipes and geothermal heat pumps can be extracted for space heating in buildings, and in some cases, pre-heating domestic hot water - by utilising ground-sourced heat the requirement for other fossil energy is reduced and therefore contributes to saving in carbon emissions – so why is the government paying mind boggling sums of money to wind farms when they are not producing electricity due to high winds – in a sane world, this money should go to encouraging geothermal heating systems.

A geothermal heating and cooling system is comprised of two distinct parts: a heat exchange unit that is installed in the interior of the home and an earth loop that is directly buried in the ground near the home, or in a body of water located by the home. The earth hose loop is made up of high-density polyethylene which is very durable and able to stand up to the elements it will encounter. This flexible hosing is installed either vertically, horizontally, or in a coiled fashion to absorb or disperse heat through the medium it circulates - geothermal heating is achieved when the system takes heat from the ground or water source and transfers it into the home through the heat exchanger - unit controls must be set to heat to achieve this. Likewise, when the switch is in the cooling mode, the system takes heat from the inside of the house and transfers it to the outside through the hose lines. As the fluid passes through the lines and the cooler ground or water, it returns to the heat exchanger to cool the home. The system uses a mixture of water and antifreeze (much like a motor vehicle radiator) to heat and cool the home.

Benefits of a Geothermal System.

- ❖ If the system loops are professionally installed, they are guaranteed for more than 50 years of trouble-free use.

- ❖ The units are very reliable and many have been in use for 30 years - all across the country many people already are enjoying the benefits of the system.

- ❖ The systems are safe to run because there are no open flames and therefore no danger of fires or carbon monoxide poisoning.

- ❖ Contributes to a healthy home, as those who suffer from dry, irritated nose or throat, will benefit as the units provide for a constant controlled humidity and temperature throughout the year.

- ❖ Simple operation allows switching from heating to cooling with a flip of a switch.

> Fuel Cells

Another option the government should be supporting and subsidising instead of ineffective wind technology is that of the fuel cell. With reference to the previous chapter Ceres Power Ltd and based in Crawley has developed a revolutionary fuel cell – the world's first commercial metal-supported solid oxide fuel cell - it starts producing energy extremely quickly and is designed to work with a range of fuels, including LPG, natural gas, methanol, hydrogen and vehicle fuels. Thus the combination of traits makes it ideally suited for mass market use - a fuel cell that offers durability, efficiency and fast response, and all at a lower cost. Indeed Ceres Power claim that their Steel Cell is a very efficient way of generating power from gas and can use the existing gas infrastructure - with their technology the cell enables up to 20 per cent more electricity for every unit of gas compared to the grid - this means that regular users could reduce the carbon footprint of their home by 25 per cent making the Steel Cell an efficient, cost effective and cleaner way of giving people control over their energy supply. A fuel cell is the most efficient way of converting fuel energy into electricity - it doesn't matter whether the fuel is natural gas or hydrogen. Fuel cells convert fuel and air directly into power and heat in a chemical reaction. This makes the process efficient, reliable and quiet. Fuel passes over the anode side and air passes over the cathode. Sandwiched between the anode and cathode is the very thin, electrolyte layer. An external circuit connects the anode to the cathode and provides the mechanism to take power from the fuel cell to power electrical devices. On their website (www.cerespower.com) Ceres Power Ltd say a single cell can power a low-energy light bulb - approximately 100 cells are combined to create a stack. One stack could supply up to 90 per cent of a home's electricity needs and all of its hot-water. The Steel Cell is completely scalable, for example 200 stacks can supply a large office, apartment block or supermarket.

No doubt the reader will have concluded that a combination of cheap shale gas and an efficient fuel cell would be a dream come true, and will wonder why government is so mentally challenged in the field of energy provision.

> Nuclear Fusion

It is recognised that the ultimate prize for generating large amounts of clean and cheap electrical energy has to be nuclear fusion – the power that drives the Sun. Conventional fission reactors harness the energy produced when atoms of uranium split, but fusion creates greater amounts of energy by fusing atoms of hydrogen – a fusion reactor's fuel is heavy hydrogen – atoms that contain one proton and one or two neutrons, and can be refined from sea water. The helium gas created when the atoms fuse is not radioactive and is harmless. Critics opine that a practical nuclear fusion power station is at least 50 years in the future, but at the Centre for Fusion Energy at Culham, near Oxford, the Joint European Torus (JET) generated (by nuclear fusion) an output of 16 megawatts back as far as 1997 – thus I would argue that given sufficient government backing and money a practical nuclear fusion power station of say 500 MW capacity is achievable sooner than the sceptics claim – remember the opinions of critics to heavier-than-air machines – a good job the Wright brothers had the will and drive to succeed – to be sure, all that is needed are people of vision, drive and determination.

> Home Insulation

Regarding future reduction in UK household electrical consumption, government should not only endeavour to more effectively encourage home insulation, but make money much more readily available to help householders in achieving this aim – all the money wasted on ineffective wind technology would have better been spent in this way – why pay millions of pounds to wind farms that are not producing electrical energy when the wind is not favourable – this is the economics of the mad house – millions could have been given to householders to help insulate their homes which would not only have reduced the amount of energy to heat their homes, but would have continued to do so at no more cost – a continuing saving in energy usage.

> Energy Saving

Great savings can be achieved if every means of illumination in the UK are changed to that of LED technology – a reduction of up to 90 per cent in electrical energy consumption for lighting.

Government should encourage and subsidise manufacturers to develop lower energy demanding goods such as washing machines, dish washers and heating appliances, all this would offer a major saving in the demand for electrical power from every household.

As mentioned earlier LG Electronics, a company in Slough, Berkshire has just introduced a Steam Direct Drive Washing Machine, which they claim, that apart from washing clothes and leaving them crease-free uses 35 per cent less water and 21 per cent less electricity than conventional models – hopefully many other companies will be spurred on and follow their fine example.

> Forced convection heating

When installing a heating system of any kind, and bearing in mind the ever increasing cost of energy these days, efficiency has to be one of the biggest factors that must be addressed. Additionally, as we continue to move towards more sensible renewable technology as an alternative source of heat provision, it has never been more important that careful thought is given to the appliances installed to deliver heat into a room or space.

There are fundamentally two ways to heat a room, either by natural convection or forced convection. Radiators, under-floor heating and perimeter heating use natural convection and fan convectors (and flame-effect fan convectors) use forced-air convection.

An enterprising company that employs fan convector technology namely, Smith's Environmental Products Ltd, Chelmsford, (www.smiths-env.com), claim that current Standard Assessment Procedure (SAP) calculations indicate a 12 per cent savings in energy when used in conjunction with a heat pump running at 40 ^{0}C and they utilise only 5 per cent of the water content of an equivalent output conventional radiator and will therefore heat up a room much faster as well as responding quickly to constantly changing weather conditions. Fan convectors and flame-effect fan convectors work efficiently within central heating systems regardless of whether they are connected to a typical boiler or renewable technology such as ground and air source heat pumps. Fan convectors and flame-effect fan convectors include a small fan so the heat can be quickly distributed around the room to give a more even temperature spread. Unlike conventional radiators which are hot to touch, fan convectors and flame-effect fan convectors have very low surface temperatures, making then completely safe and therefore ideal for children and the elderly. Fan convectors are considerably smaller than the equivalent conventional radiators and are more versatile. They are designed to install in those 'dead' spaces, so you don't have to design the room around the heating. Fan convectors require connection to a 'wet' central heating system and a mains electrical connection to run the fan.

Smith's Environmental Products say they are thoroughly committed to recycling and waste management. All processes incorporate waste minimisation procedures ensuring minimal waste and maximum recycling. All their products are designed with energy efficient performance in mind, intended to work from both existing and renewable sources of energy. All the product packaging is fully recyclable, offering further testament to their commitment to the environment. – the companies quality assurance management system has been awarded ISO 9001: 2008 Quality Management Systems certification – an internationally recognised benchmark for quality. With only around 8 per cent of UK businesses achieving this prestigious award, Smith's Environmental Products is at the forefront of quality service and customer care.

Therefore, apart from a new build, it would make both economic and environmental sense to evaluate the cost of upgrading any existing conventional heating system to that of a forced convection heating system.

In the next chapter the author shows how he is currently saving over £1100 per year on household energy consumption without any loss in creature comforts.

"How, sir, would you make a ship sail against the wind and currents by lighting a bonfire under her deck? I pray you, excuse me, I have not the time to listen to such nonsense."

Napoleon Bonaparte, when told of Robert Fulton's steamboat, 1800s

CHAPTER SEVEN

How to Reduce your Domestic Energy Bill

With the ever increasing cost for the heating and lighting of our homes then any effort made to contain such expenditure by more efficient home insulation in the form of loft insulation, cavity wall insulation and double glazing, coupled with switching off lights in empty rooms, and turning off electrical/electronic gadgets and equipment when not needed, should be applauded. Regarding the insulation of the home and consideration on the return of capital spent, then loft insulation should be carried out first, followed by cavity wall insulation and then double glazing – needless to say that draughts, such as under doors, should be dealt with straight away: home insulation is covered in more detail in the Summary section at the end of this chapter.

If the home dweller can minimise their demand for resources, such as fossil fuels, in producing heating and lighting in the home then so much the better – the more efficient use in white goods, lighting and heating will conserve Earth resources, whilst at the same time reducing household costs. This chapter shows how to reduce electrical consumption and at the same time reducing the amount of say, natural gas or oil, for home heating.

These savings are easily attainable - not only have I been able to achieve savings in heating (reduction in LPG consumption) and lighting costs, but to actually cut the electricity bill significantly when living in Pembrokeshire, Wales, and I would stress this was achieved at no loss to creature comforts.

When my wife and I lived in a four bedroomed detached house in Pembrokeshire we managed to reduce our electricity bill from an average of £500 per annum to £294.4 per annum, a reduction of 41.12 per cent. The electricity savings were achieved by the employment of low energy lighting, but predominantly from the installation of a 2.3 kW solar panel system, which also helped in achieving a reduction in the amount of natural gas (LPG) used for cooking *and* heating - it should be noted though that if all the exported energy (see Solar iBoost later) had been used by the household then the saving would have amounted to over 50 per cent.

This chapter will clearly demonstrate to the reader how they can reduce not only the household electricity bill – cutting it in half or more - but also in the reduction of other household energy consumption such as gas, oil or solid fuel, whilst still maintaining, or indeed, improving living standards. To this end the chapter illustrates the actual savings made by the author whilst living in Pembrokeshire – but especially more so, as we shall see later, at the property in Ceredigion, Wales.

Pembrokeshire Savings

This first example entails a four bedroomed detached house where my wife and I lived from 1999 until 2013, and considers energy savings from the employment of low energy lighting and the installation of a solar PV system - for further detail see Appendix 3.

A solar power array was installed on the roof of the detached garage with the panels having a tilt of approximately 35 degrees and a southerly orientation. The solar array consisted of ten 230 watt Dimplex high performance polycrystalline solar PV modules, offering a total system capacity of 2.3 kW. The capital cost of the installation was £7,818.

The mains electrical power consumption before the installation of the solar panels was of the order of 5000 kWh per annum - this level of consumption was a result of employing a mix of incandescent, CFL and LED bulbs for lighting – efforts were also made to purchase (when necessary) low energy demanding white goods.

Over the evaluation period (2012) rain and floods occurred during the early part of the year, followed by a sunny March and April - but this was followed by a hopeless summer. Generally the weather was quite dismal for 2012, as is reflected in the solar generation shown in the table below. It was telling that the solar generation for May was greater than June by 65.5 kWh where it would be expected the greater generation would occur during June, as this month apart from longer days (summer solstice this month), can be very sunny with the strongest sunlight. Therefore 2012 was not considered an average year in the context of sunshine hours.

But with everything considered it was concluded the solar array performed better than anticipated, see table below:

Pembrokeshire Solar Generation for 2012, figures in kWh

JAN	FEB	MAR	APR	MAY	JUN	JUL	AUG	SEP	OCT	NOV	DEC	TOTAL
42.5	93.1	219.4	181.8	306.9	241.3	257.1	230.4	219.6	137.7	83.2	43.2	2056.2

As mentioned earlier the mean annual power consumption from the local power network from 1999 to 2013 was of the order of 5000 kWh per year. Thus the solar generation (2056.2 kWh) during 2012 turned out to be 41.12 per cent of the mains consumption, and this being achieved during a predominantly cloudy year.

Feed In Tariff (FIT)

The Feed in Tariff allows anyone who generates their own electricity, the opportunity to earn an income for 20 years and thus cutting their energy bills. Feed in Tariff income is tax free for all domestic owner occupiers, and is Index Linked from April each year.

There are three ways to earn and save money from Solar PV Feed in Tariffs:

1. **The Generation Tariff.** This pays for each unit (kWh) of electricity that is produced, regardless of whether or not it is used by the consumer. At the time of writing the FIT pays 14.9 pence per unit of electricity generated for systems up to 4 kW.

2. **Export Tariff.** For installations without an export meter the Export Tariff pays, at the time of writing, an additional 4.64 pence for each unit (kWh) of electricity that is not used and is exported back to the National Grid – the amount of energy

exported is deemed to be 50 per cent of that generated. Thus, for example, if the system generated 4000 kWh over a period of time, then 2000 kWh is considered to have been exported over the same period of time.

3. **Electricity Bills.** Substantial bill savings are to be gained if the electricity that is generated is consumed at source - a consequence of fewer units (kWh) on the power bill.

Solar Energy Earnings Pembrokeshire

When generation began for the house in Pembrokeshire the FIT payment was a generous 43.3 pence per kWh generated and it was deemed, by the electricity supply company that 50 per cent would be fed back (exported) into the local network at a payment of 3.1 pence per kWh. During 2012 a total of £922.2 was paid for solar energy generated, although it should be noted that for simplicity sake I have kept to the initial value of the FIT as due to index linking (April) the value of the FIT increased to 45.4 pence for each unit generated and 3.2 pence for each unit exported – using these figures would make the earnings marginally greater.

Electricity Bill Savings Pembrokeshire

It should be noted that not all the solar generated electricity was consumed in the house in Pembrokeshire, as a certain amount of the solar generated power had in fact been exported to the local power network. A good example being when export took place on a sunny day in the summer, when the occupants have spent the day at the seaside, leaving the house empty and thus a minimal need for electrical energy - low energy demanding items such as fridges and freezers - the property in Pembrokeshire contained a fridge in the kitchen and two freezers in the garage. Bearing in mind that at other times of the year as during the winter months, when solar generation is low, most if not all, solar generation was used at source. For simplicity sake the electricity company Standing Charge is not considered.

Not knowing exactly how much solar generated electricity was exported, it is assumed that all solar generation was used - thus the electricity bill for 2012 amounted to £294.4 instead of the average annual £500 offering a saving of £205.6 - in addition £922.2 was paid for solar generation under the FIT scheme.

Total Income Pembrokeshire

If both earnings and savings are combined and defined as income then the total income for the year amounted to £205.6 + £922.2 = **£1127.8**. Projecting this income over 20 years would amount to **£22,556**.

The total income over 20 years has to be corrected by deducting the initial capital cost of £7818 for the solar PV system which then gives a total of (22,556 – 7818) = £14,738 which is deemed very acceptable both in energy conservation and as an investment – where could you get such a return on your money these days with interest rates at their lowest for years.

It should also be noted that with a saving of £1127.8 per annum the initial capital cost of the solar PV system is recoverable in 6.9 years.

The house in Pembrokeshire depended on natural gas (LPG) for cooking and central heating and apart from the savings in electrical energy during 2012 we noted a reduction in gas usage as the house heating was complimented by the careful use of electrical heaters – the conclusion being that actual export was much less than 50 per cent during the year.

When the solar PV system was purchased installation costs were high - but the energy savings and the return on capital were proven to be very attractive due to the fact that FIT payments were very good.

Today, although the FIT payments have reduced, system installation is that much cheaper having reduced in price by about 50 per cent and the reader should now be able to have a 2.5 kW system installed for less than £4,000 – indeed, I have seen adverts for 4kW capacity systems priced just under £5000 - therefore the installation of a solar PV system is still deemed a very attractive proposition, both from the earnings of the FIT, and of course, the saving in electricity and other energy costs.

The second example involves a five bedroomed detached dwelling and considers not only a solar PV system and LED lighting, but air-to-air heating, voltage regulation and a Solar iBoost unit that has the ability to identify and enable the consumption of 'exported solar energy' at source.

Five Bedroom Detached Property, Ceredigion

During 2013 my wife and I moved to a five bedroomed detached property in Ceredigion and with the experience gained in Pembrokeshire from the use of low energy lighting, solar PV generation, and a preference for low energy white goods, it was decided to fully embrace as much energy saving technology as possible within our budget and the constraints of the detached home we moved to in Ceredigion – such constraints as an existing ten year old oil boiler for hot water and central heating. If we were building a house from scratch then consideration would be given to geothermal heat energy in conjunction with a solar PV system, air-to-water pump (more efficient than an air-to-air heat pump), and a voltage regulator, all monitored and controlled electronically to minimise energy usage - all lighting would be by LED technology, and all white goods would be of the low energy type.

With the correct approach and a basic understanding of electrical power most property owners should have little difficulty in reducing their energy consumption without any reduction to their quality of life – something every householder would wish to achieve in this age of ever increasing energy costs.

Thus the following will show that even greater savings can be achieved in energy usage complimented with a decrease in energy bills by employing not only solar PV panels and LED lighting, but air-to-air heating, voltage regulation and an electronic unit known as a SOLAR iBOOST which has the ability to identify potential 'exported solar energy' and intelligently control and adjust the flow of energy to the immersion heater instead of the local power network.

The solar panels were mounted on the property roof with a tilt of approximately 40 degrees and a south westerly orientation. The solar array consists of sixteen Hyundai 250 Watt Black Frame Polycrystalline panels with a total system capacity of 4 kW at a cost of £7,100.

The annual mains power consumption, without solar, was historically assessed at 6000 kWh – this assessment is based on a five bedroom dwelling and the use of incandescent light bulbs. The house heating is by oil (underfloor heating on the ground floor and radiators upstairs) with an annual oil consumption of 1500 – 2500 litres depending on how cold the weather happened to be.

Cooking is by an electrical and gas (LPG) stove. LPG gas consumption approximately 38 kg per annum - it should be noted that some backup heating is available from a wall fitted gas fire in the lounge during extremely cold weather.

We have seen from the table of solar generation for Pembrokeshire during 2012, the solar generation for May was greater than June, where it would be expected that the greater generation would occur during June – but 2012 turned out not to be a typical year as the generation recorded for Ceredigion has confirmed – the maximum generation during June was closely followed by July then August.

Ceredigion Solar Generation (kWh) for 2014

JAN	FEB	MAR	APR	MAY	JUN	JUL	AUG	SEP	OCT	NOV	DEC	TOTAL
88	197	331	424	471	658	567	486	424	219	166	100	4131

Solar Energy Earnings Ceredigion

Similar to the Pembrokeshire calculations, the value of the FIT payment for Ceredigion at the start of generation in January, were used for the whole of the year and in this case amounted to 14.9 pence per unit (kWh) generated and 4.64 pence for each unit exported. Again it was deemed by the electricity supply company that 50 per cent of solar generation would be exported back into the local network. During 2014 a total of 4131 kWh was generated by the solar panels and the total TIF payment amounted to £711. Projecting the 2014 savings earnings over 20 years (ignoring Index Linking for simplicity) then a total of £14,220 would be earned, see Appendix 3.

Electricity Bill Savings Ceredigion

The annual consumption of mains electricity for 2014 was 4200 kWh – a saving of 1800 kWh (30 per cent) on an assessed annual usage of 6000 kWh, see Appendix 3 for more detail.

This amounted to an electricity bill saving of £221

The annual cost of mains electricity was £515.76 with the cost per unit (kWh) at the time of 12.28 pence - again the company Standing Charge has been ignored for simplicity sake.

Projecting this figure over 20 years would result in £10,315.2 being saved.

Total Income Ceredigion

Again if both earnings and savings are combined and defined as income then the total income for the year amounted to £711 + £221 = **£932**

Projecting this income over 20 years would amount to **£18,640.**

Again the above total income over 20 years has to be corrected by deducting the initial capital cost of £7,100 for the solar PV system which then gives a total income of £11,540, which again cannot be bad in both an attempt to being 'green' and as an investment. Again where would you get such a return on your money these days with current interest rates?

With an income of £932 per annum the initial capital cost of the solar PV system is recovered in 7.6 years.

It is interesting that the projected payback time for the solar PV system for Pembrokeshire is 6.9 years whilst Ceredigion at 7.6 years, with a capital difference of £718 between the installation costs of the systems, see Appendix 3.

Although the real surprise comes in the earnings whereby the house in Pembrokeshire earned £4,224 more than the property in Ceredigion over a 20 year period even though Ceredigion (4 kW) generation capacity is nearly double that of Pembrokeshire (2.3 kW) generation capacity.

This is due to the difference in the FIT generation payment of 30 pence per kWh between the two examples:

Pembrokeshire = (20 x 2056.2 x 0.433) + (20 x 2056.2 x 0.031 x 0.5) = £18,444

Ceredigion = (20 x 4131 x 0.149) + (20 x 4131 x 0.0464 x 0.5) = £14,227

But this is not the full story for Ceredigion - greater savings from solar PV generation are now being achieved by the installation of a unit called a Solar iBoost, that has ability to identify potential 'exported solar energy' and intelligently control and adjust the flow of energy to the immersion heater instead of the local power network – this unit is manufactured by a UK company known as Marlec Engineering Co Ltd (www.marlec.co.uk) – which has in the six month period from the beginning of October 2014 up until the end of March, 2015, diverted 370 kWh of electricity to the property hot water immersion heater. This energy saving is equivalent to running a 2 kW electric fire for 185 hours, or if you like, running the electric fire for 4 hours each day for just over 1.5 months – and it is not rocket science to work out that the summer months will provide far greater savings in electricity and indeed, oil usage for heating hot water.

Therefore we will now look at this innovative energy saving Solar iBoost unit in greater detail – every household that has solar PV panels will surely wish to invest in this very modestly priced option.

Solar iBoost

Most solar PV systems are installed with just a single generation meter which does do not differentiate between generation used at source or that exported to the local power network - the installed meter can only show the total solar generation. As such it should be understood that the householder will not easily be aware of energy exported at any one moment, and from the householders point of view, will be lost to the power network – although this energy being available to another power company paying customer connected to the local power network.

This is where the Solar iBoost comes into its own as the unit has ability to identify potential 'exported solar energy' and intelligently control and adjust the flow of energy to the house immersion heater instead of the local power network. In effect this means that practically all of the power generated by a solar PV system can be used at source, much to the delight of the householder…a considerable saving over time. To fully quantify the operation of the unit it should be noted that when export levels drop below 200 watts the unit switches off and waits until the export of 200 watt or more is restored, the unit then returns (switches) to water heating.

The author, as mentioned above, had a Solar iBoost unit installed at the beginning of October, 2014 and at March, 2015 it has diverted 370 kWh of electrical energy to the hot water tank immersion heater, thereby saving in the use of oil for heating the house hot water – noting the saving, as mentioned earlier, occurred during the winter months – so there will obviously be much more electrical energy to divert during the summer months.

The benefits of the Solar iBoost are summarised as follows:

Solar iBoost fits quickly and neatly into an airing cupboard, simply wired between the existing fused spur and the immersion heater.

- ❖ It wirelessly receives information continuously from a sender device which activates the Solar iBoost to start water heating.

- ❖ The sender is battery powered so it is rapidly installed with its clamp in the utility meter box and no need for expensive wiring.

- ❖ The Solar iBoost intelligently controls and adjust the flow of energy to the immersion heater as the home consumption varies ensuring that only excess power is used.

- ❖ There is no need to change the immersion heater as the Solar iBoost works with any normal household immersion rated up to 3 kW.

- ❖ Solar iBoost displays real time and historical energy savings figures and LED symbols indicate the operating status.

- ❖ Simple timer programming enables Solar iBoost to work in harmony with Economy utility tariffs and a boost override switch means you can top up from the Grid any time extra hot water is needed.

Electrical energy diverted from the Solar iBoost does not affect FIT payments - for householders with a deemed usage contract from their FIT provider using Solar iBoost means a greater or full usage of the free energy and still receive the 50 per cent export payment. When an export meter is fitted the benefits of the Solar iBoost can still outweigh the rising costs of water heating.

The Solar iBoost was obtained from, Marlec Engineering Co Ltd, Rutland House, Trevithick Road, Corby, Northants, NN17 5XY. (www.marlec.co.uk), telephone: 01536 201588.

At the time of writing the Solar iBoost unit could be purchased for just under £250.00 which I consider excellent value and would recommend any household with a Solar PV array and hot water tank to have a unit installed – a no brainer as they say.

To further enhance the energy savings from the solar PV system it was decided to invest in an air-to-air heat pump – the heat pump would use mains electricity far more economically in heating the house and, in effect, run free of cost when using solar generated electricity – what's there not to like.

For those not familiar with air source heat pumps the following will hopefully explain the technology in more detail and enlighten the reader to their potential for reducing energy costs.

Air Source Heat Pump

Fundamentally an air source heat pump is a system which transfers heat from outside to inside a building, or vice versa. For domestic heating purposes an air source heat pump absorbs heat from external air and releases the heat energy inside the building, as either hot air, hot water-filled radiators, underfloor heating or domestic hot water supply; the same system can often do the reverse in summer, cooling the inside of the house.

The heating performance and efficiency of an air source heat pump system is commonly measured by the Coefficient of Performance (CoP). The CoP is a simple calculation which works out how much energy the heat pump is able to extract from the energy source compared to the amount of electrical energy it uses.

CoP =   Heat output of system (useful heat)
        Electrical input from compressor and fan motors

For example: 6 kW heat pump
                   1.2 kW of electrical input   = CoP of 5

Generally speaking, the higher the CoP figure, the greater the efficiency of the heat pump. A CoP however only applies to a specific temperature, which means that the CoP rating is not representative of the performance that could be achieved across a whole year. A far more accurate assessment of efficiency therefore is provided by the Seasonal Coefficient of Performance (SCOP). It defines the performance of the heat pump over the course of the year, with seasonal variations in conditions.

Air-to-Air Heat Pump

As recommended by our solar panel installer a Greensource air-to-air heat pump was installed at the end of January 2014 and has up until the end of January, 2015 (12 months) saved 300 litres of heating oil. At a price of 54.66 pence per litre of oil (2013 prices) a saving of £164 has been achieved, not forgetting the significant electricity saved and a much warmer house.

According to Worcester (Bosch Group) their Greensource air-to-air heat pump offers highly efficient heating in all seasons, with an industry leading SCOP of 3.8. Worcester also state there is a maximum heating output of up to 6 kW, which is sufficient for heating an area of up to 100 m^2 and it is this performance coupled with the highest levels of quality and reliability which are synonymous with the Bosch name throughout the world that persuaded me to purchase the Greensource air-to-air heat pump.

Apart from allowing consumers to efficiently heat their homes, the Greensource air-to-air heat pump also offers a cooling feature which ideally complements PV systems on a hot summer day when solar generation is at a maximum, with the house inside temperature correspondingly high - effectively it offers air conditioning for free – and you cannot get better than that!

The Worcester Greensource air-to-air heat pump is suitable for a wide variety of property types and sizes, and can complement existing gas, oil or renewable hot water systems. It can also offer stand-alone heating for the home in certain applications. The total cost for purchasing the heat pump and installation was £1991.85 which the author hopes to recuperate over a very short time due to the savings in both oil and electricity for a given amount of heat energy. I am very pleased with the installation of the air-to-air heat pump as it is a brilliant concept especially when used in conjunction with solar PV panels - I should point out to the reader that the Greensource air-to-air heat pump did not have a weekly programmer built-in, although it did have a 24 hour programmer offering timer on and timer off - there are many other makers and suppliers of air-to-air pumps on the market, and maybe one of these other options might have a weekly programmer built in, should the reader desire this feature.

Voltage Regulator

Living in a rural environment with predominantly overhead power distribution a variation in the level of voltage was soon noted at Ceredigion which gave concern regarding stress to sensitive electronic components in not only items such as the TV, radio and music players but also to computer equipment. It was recognised that by ensuring a correct (lower) voltage supplied to the house then a saving in power would also be achievable (remember power = voltage x current). Thus on this basis a PowerSines Smart HS-100 unit was installed at a cost of £599 during December, 2013, neatly next to the consumer unit in the garage.

PowerSines state in their literature that using their patented RIGHTVoltage technology, they can ensure that just the right amount of energy needed to maximise efficiency is delivered to the various electrical appliances the consumer is using. The optimal voltage for electrical equipment is 220 volts; however the typical UK incoming voltage averages at 242 volts or more. This wasted energy is very costly and harmful to electrical equipment lifespan.

PowerSines sum up the benefits (with acknowledgements) as follows:

- ❖ Installs without any changes to the existing infrastructure.
- ❖ Provides smart metering of energy consumption.
- ❖ Handles all household loads.
- ❖ Seamlessly integrates with PV systems.
- ❖ Extends electrical equipment lifespan.
- ❖ Installs quickly and simply.
- ❖ Small footprint.
- ❖ Includes under-voltage protection.
- ❖ Ensures voltage optimisation for the entire property.

The company also claim the following savings in a typical household,

- ❖ Lighting = 22 per cent.
- ❖ Generic Power Outlets = 7 per cent.
- ❖ Heating, ventilation and Aircon = 12 per cent.
- ❖ Kitchen appliances = 14 per cent.

Summary

The greatest energy savings were achieved for the property in Ceredigion with the resultant annual savings and the means of which, are open to all householders and are summarised as is as follows.

✓ Home Insulation

In conserving the amount of energy consumed in a household the first steps are to ensure the building is well insulated to contain heat from the Sun and that of costly commercial heat – a bit pointless to turn up the heat and having it escape through ill-fitted doors and windows, uninsulated roof spaces, walls and single glazing – just like throwing money out of the window.

Therefore in order of cost and effectiveness all leaks from badly fitted doors and windows should be eliminated first. Next attention should be given to roof spaces and a layer of insulation between 120mm-270mm should be laid down between the roof space floor joists – to improve insulation further then lay rigid insulation boards on top, with wooden boarding on top of that – it is possible to buy insulation board pre-bonded to floor boarding to make the job easier. A typical saving for a detached house is £25 per annum with a typical installation cost of £310. To ascertain how much can be saved then visit the Energy Saving Trust (www.energysavingtrust.org.uk).

Having adequately insulated the roof space then attention should be given to cavity-wall insulation - the average 3-bedroom house cavity wall insulation job should cost around £250 with subsidies, £500 without subsidies. The energy savings are significant with a saving of about £115 per year on heating - more information is available from the Energy Saving Trust as mentioned above.

Older buildings with solid walls basically have two choices, insulate from the inside or the outside - insulating solid walls can cut heating costs considerably - solid walls let through twice as much heat as cavity walls do, and will cost more than insulating a standard cavity wall, although the savings on heating bills will be more. Internal wall insulation is done by fitting rigid insulation boards to the wall, or by building a stud wall filled in with insulation material such as mineral wool fibre. External wall insulation involves fixing a layer of insulation material to the wall, then covering it with a special type of render (plasterwork) or cladding. The finish can be smooth, textured, painted, tiled, panelled, pebble-dashed, or finished with brick slips. A typical cost of insulating an internal wall will be in the region £3,000 - £14,000 and that for an external wall, £5,000 - £18,000, and so not cheap but will save about £455 per annum on heating costs for a detached house - again more information is available from the Energy Saving Trust.

✓ Low Energy Lighting

Having insulated the house adequately, as in the case for both the house in Pembrokeshire and the dwelling in Ceredigion, the next consideration is the fitting low energy lighting. Whilst living at the house in Pembrokeshire incandescent bulbs were still widely used and the next best alternatives were that of CFL technology. But the greatest saving came from fitting LED bulbs in Ceredigion saving up to 90 per cent in energy use. Taking the kitchen at the property in Ceredigion, as an example, nine 50 watt halogen (GU10 fitting) lights were changed for nine 5 watt MiniSun LED (GU10 fitting) lights which gave a saving of 405 watts at any one instance for illuminating the kitchen - all other lights were changed to LED bulbs giving a total saving of 90 per cent for the whole property.

✓ Solar PV Panels

The earnings and savings (total income) for 2014 for the property in Ceredigion amounted to £932 and using this figure to project total income over twenty years would amount to £18,640 – deducting the capital outlay of £7100 leaves a sum of £11,540 - thus apart from all the other attributes such as energy saving and helping to contain emissions, it was deemed a very good investment on capital spent. The reader should note that the solar panels mounted on the roof in Ceredigion were not installed in the optimum position due to the orientation of the property. The actual roof space that offered the most exposure to the Sun, and to which the solar panels were attached, had an orientation towards the south-west. It was observed that generation on a sunny day did not enter the kilowatt range until after 1100 hours whereas if the panels were mounted facing more to the south then higher generation would have occurred earlier in the day. The ultimate solar array would track the Sun during the day, but this would possibly require a ground installed system, which would be impractical for many households, not to mention high installation costs thus making the whole concept both impracticable and uneconomical. The best the householder can go for is a suitable roof space compromise - a north facing roof would obviously be the worst possible siting, and should not be considered.

✓ Air Pumps

The installation of an air-to-air pump was another leap forward in saving energy and recognising that when the Solar panels were producing electricity the air-to-air pump was running (putting the capital cost to one side) without *any cost* – and you cannot get better than that – heated air in the winter, and cool air in the summer. Over a twelve month period the air-to-air unit has saved 300 litres of heating oil. At a price of 54.66 pence per litre of oil (2013 prices) a saving of £164 has been achieved, not forgetting the significant electricity and other energy saving and a much warmer house.

✓ Solar iBoost

When I first came across this remarkable little unit I was very impressed as it offered a means to direct 'export' solar generation back into the household without affecting the tariff payments – naturally there was no hesitation in having a unit installed 'asap'. As mentioned earlier the unit was installed at the beginning of October 2014 and during a six month (winter) period, diverted 370 kWh of electrical energy to the hot water tank immersion heater. The subsequent cost saved for electricity at 12.28 pence per unit amounted to £45.44 plus a saving in the use of oil for heating the house hot water – this saving, occurring during the winter months. Obviously there will be much more electrical energy to divert during the summer months and thus greater savings. Luckily the property at Ceredigion had the emersion heater element and thermostat situated at the base of the hot water tank which enabled the whole tank to be fully heated. A lot of hot water tanks have the emersion element installed either half way up or from the top of the tank with the thermostat situated half way up the tank – thus consideration should be given to modification such as moving the thermostat to the bottom of the tank and adjusting accordingly, and for top installed elements which are short in length, to be changed for a much longer element – care should be taken in any modification to avoid excessive heat and boiling the water in the tank.

- ✓ Voltage Regulator

The savings from the Voltage regulator are yet to be fully quantified, suffice to say that the various percentage savings by the manufacture have been accepted in conjunction with the claim of its ability to extend the lifespan of electrical equipment by maintaining a constant and optimising voltage for the entire property.

The annual energy saving for Ceredigion for 2014 amounted to

£932 + £164 + £45.44

= £1141.44

So what's there not to like

It can easily be seen from the property in Ceredigion the 'earnings and savings' almost amounted to free space heating. No doubt the reader will have also appreciated the total energy savings will be far greater in smaller dwellings where less heating will be required – indeed a lot of folk could easily realise greater savings again than has currently been achieved for the property in Ceredigion.

The reader should note that the electricity savings from the employment of the air-to-air pump have not been factored into the total savings. The reason being due to the difficulty in actually measuring the electricity saved, suffice to say, the manufacturers claimed savings appeared very reasonable, especially when backed by the good name of Bosch. Thus I would assess the savings in electricity and oil would be greater than £164 and in excess of £200, thereby offering a more realistic total saving for the property at Ceredigion of at least **£1200.**

I would recommend anyone about to plan and build a new property to consider a fully integrated system which would consist of, geothermal heating, both electrical (PV) and thermal solar panels, air-to-water (more efficient than air-to-air, and similar to geothermal, both should be eligible for the Renewable Heat Incentive, unlike air-to-air systems), Solar iBoost and a voltage regulator. The integrated system to take full advantage of LED technology with the widespread use of LED lighting, and with such a system all heating and lighting would offer minimum or no running costs once capital cost has been discounted.

Finally, as mentioned in the previous chapter, where a household has existing conventional radiators then it would make economic and environmental sense to assess the cost of upgrading to forced convection radiators and possibly a flame-effect fan convector.

Appendix 1

Wind

Calculations given in SI (System International) units.

kW = 1000 watts
MW = 1000,000 watts (1,000 kW)
GW = 1000, 000, 000 watts (1,000 MW)
TW = 1000, 000, 000, 000 watts (1,000 GW)

Unit of electrical power = 1 kW consumed over 1 hour = 1 kWh.

Power available for wind generation

Now power is the rate at which energy is available – or the rate at which energy passes through an area per unit of time, therefore power:

$P = \frac{1}{2} dav^3$

If the value for air density at sea level is substituted for d in the above equation then power in watts:

$P = 0.6125 av^3$

Where area is in square metres and v is in metres per second, or:

$P = 0.0508a\ v^3$

Where area is in square feet and speed is in miles per hour.

Therefore for a wind speed of 5.4 metres per second (12 miles per hour), a wind generator with a propeller spanning 2.4 metres (8 ft) in diameter will produce the following power:

$A = \pi r^2 = 3.142 \times (2.4 \times 0.5)^2 = 4.5$ square metres

Thus,

$P = 0.6125 \times 4.5 \times (5.4)^3$

   $= 2.760 \times 157.46$

= 434.6 watts

But if we double the wind speed to 10.8 metres per second we have:

$P = 2.760 \times (10.8)^3$
  = 3476.8 watts

This, as we can see, provides eight times as much power.

Swept Area

In the next example we will determine how much power can be produced from increasing the propeller (rotor) span from 2.4 metres (8ft) to 3.4 metres (11 feet) for the same wind speed of 5.4 metres per second (12 miles per hour).

Area = $3.142 \times (3.4 \times 0.5)^2 = 9$ square metres

(*Note: we have doubled the area swept from 4.5 square metres to 9 square metres*)

Thus,

$P = 0.6125 \times 9 \times (5.4)^3$
  = 5.5 × 157.46

  = 868 watts

For the same wind speed, by increasing the propeller diameter from 2.4 to 3.4 metres, (doubling the area [a] swept), we have a gain in wind power of 433.4 watts, virtually doubling the wind power.

Increase in wind speed and area swept

Now let us see what happens if we double the wind speed from 5.4 (12 miles per hour) to 10.8 (24 miles per hour) when the propeller diameter is 7 metres:

$P = 23.58 \times (10.8)^3$

  = 23.58 × 1259.7

= 29,703.7 watts

Again we can see that a doubling of the wind speed coupled with a substantial increase in propeller diameter will result in a significant increase of power to almost 30 kilowatts.

Wind generator height above ground level

When choosing a site for a wind generator it is important to recognise that obstructions on or near the ground disrupt the flow of wind – increasing the height will have a marked effect on wind speed and power. One way to calculate the increase in wind speed with height is to employ what is known as the 'power law' equation:

$$V = (H/Ho)^\alpha \, Vo$$

Where V = wind speed at new height
Vo = wind speed at original height
H = new height
Ho = original height
α = the surface rough exponent

Now the rate at which the speed of the wind increases with height varies with the surrounding vegetation and terrain. For example, the increase is greatest over rough terrain or numerous obstacles such as trees and shrubs, but lowest is for surfaces such as smooth water or ice - the surface roughness exponent (α) can vary from 0.1 for ice and water to 0.28 for woodlands and suburbs. Let us assume our wind generator again has a propeller with a span of 2.4 metres (8ft) and is subjected to a wind speed of 5.4 metres a second (12 miles per hour). We will also assume the wind generator is supported originally by a tower 10 metres tall situated in an area of low grass (α = 0.14) – then we change the tower to one that is 50 metres tall:

$V = (50/10)^{0.14} \times 5.4$ (note $5^{0.14}$ *is not simple to calculate and has been arrived at by use of tables*)

= 1.25 x 5.4

= 6.75 metres per second

By increasing the height from 10 to 50 metres we have increased the wind speed by 25 per cent. But the power increases more dramatically because of its cubic relationship with speed and therefore we have:

$P = 0.6125 \times 4.5 \times (6.75)^3$
$= 2.760 \times 307.5$
$= 848.7$ watts

Therefore by increasing the height by a factor of five nearly doubles the power available.

Appendix 2

SI (System International) units.

kW = 1000 watts
MW = 1000,000 watts (1,000 kW)
GW = 1000, 000, 000 watts (1,000 MW)
TW = 1000, 000, 000, 000 watts (1,000 GW)

Unit of electrical power = 1 kW consumed over 1 hour = 1 kWh.

Wind Generation

Wind Generator vs Coal-Fired Power Station

To determine the equivalent number of 2 MW, 123 metre (400 feet) high, wind generators required to replace a single typical coal-fired power station.

As an example we will take Aberthaw B coal-fired power station which has an installed capacity of 1500 MW.

If we take the load factor of a 2 MW wind generator as typically in the order of 25 per cent, and Aberthaw B having a load factor of 62 per cent (using DTI 2005 figure), then:

Potential output of Aberthaw = 1500 x 0.62 = 930 MW
Potential output of a 2 MW wind generator = 2 x 0.25 = 0.5 MW

Thus we have, 930/0.5 = 1860 wind turbines.

Therefore to replace a large coal-fired power station such as Aberthaw, would require nearly two thousand 2 MW, 123 metre (400 feet) high wind generators.

Note: DTI (Department of Trade and Industry) became the Department for Business, Enterprise & Regulatory Reform (BERR) and is now The Department of Energy and Climate Change (DECC).

To determine the equivalent number of wind generators that would be needed to compete with the following seven large coal-fired power stations:

Station	Installed Capacity (MW)
Ferrybridge C	1955
Fiddler's Ferry	1961
Eggborough	1960
Cottam	2008
West Burton	1972
Rugeley	1006
Aberthaw B	1500
Total	12362

We will again take the for load factor for a coal-fired power station as 62 per cent (DTI 2005 figure), and a load factor of 25 per cent for a large 2 MW wind generator.

Total output = 12362 x 0.62 = 7664 MW
Output of wind generator = 2 x 0.25 = 0.5 MW

Thus, 7664/0.5 = 15,328 wind generators.

Therefore to compete with seven large coal-fired power stations would require just over fifteen thousand 123 metre (400 feet) tall, 2 MW, wind generators.

Now 20 wind generators might extend over an area of 1 square kilometre - so 15,328 would require over 766 square kilometres of land – this would truly be utter madness!

Appendix 3

Domestic Energy Savings

SI (System International) units:

kW = 1000 watts
MW = 1000,000 watts (1,000 kW)
GW = 1000, 000, 000 watts (1,000 MW)
TW = 1000, 000, 000, 000 watts (1,000 GW)

Unit of electrical power = 1 kW consumed over 1 hour = 1 kWh.

Hours in a year = 24 x 365 = 8760

10^2 = 100
10^3 = 1000
10^6 = 1,000,000
10^9 = 1,000,000,000
10^{12} = 1,000,000,000,000

Before assessing the overall energy savings that can be achieved in a four or five bedroom detached property it will be useful to consider the banned filament bulb and the energy saving lighting technology that is now available for the consumer in the UK.

Incandescent Light Bulbs

An incandescent light bulb (Tungsten bulb) produces light by means of a wire filament heated to a high temperature by an electric current passing through it until it glows. The hot filament is protected from oxidation with a glass or quartz bulb that is filled with inert gas or the bulb being evacuated. In a halogen lamp, filament evaporation is prevented by a chemical process that re-deposits metal vapour onto the filament, extending its life. The light bulb is supplied with electrical current by feed-through terminals or wires embedded in the glass. Most bulbs are used in a socket which provides mechanical support and electrical connections.

Tungsten bulbs were manufactured in a wide range of sizes, light output, and voltage ratings, from 1.5 volts to about 300 volts. They require no external regulating equipment, had low manufacturing costs, and worked equally well on either alternating or direct current. As a result, the incandescent lamp was widely used in household and commercial lighting, for portable lighting such as table lamps, car headlamps, and flashlights, and for decorative and advertising lighting.

Filament bulbs are much less efficient than most other types of modern electric lighting as they convert less than 5 per cent of the energy they use into visible light, with the remaining energy being converted into heat - the luminosity of a typical incandescent bulb is 16 lumens per watt, compared to the 60 lumens per watt of a Compact Fluorescent Light bulb (CFL) - they also have short lifetimes compared with other types of lighting; around 1000 hours for filament light bulbs versus typically 10,000 hours for Compact Fluorescents (CFLs) and 30,000 hours for Light Emitting Diodes (LEDs).

Compact Fluorescent Lights (CFLs)

Under EU rulings the use of tungsten filament bulbs are now BANNED in the UK and as a result many people have replaced their conventional bulbs with low energy types employing CFL technology - although they are much cheaper to run many people are not enamoured with them as they suffer from flicker, give out a limited, cold light and cannot be used with dimmer switches; they can also be relatively large and ugly in appearance.

Nevertheless it is useful to assess the energy savings by replacing filament bulbs with CFL bulbs in households in a small country such as Wales – the small population allows for less cumbersome mathematics and easy extrapolation for countries with far greater populations.

To assess the potential energy savings, the amount of electrical energy consumed during the winter months of October to March inclusive is calculated for every council tax dwelling in Wales – assuming that during the six-month period (2005 to 2006), each dwelling employed six 20-watt CFL bulbs instead of six 100-watt tungsten filament bulbs, then six 60-watt tungsten filament bulbs.

Council tax dwellings for Wales during 2005-06 = 1,279,494
(Source: national statistics www.wales.gov.uk/statistics)
For the purpose of the exercise and to keep the maths simple 1,279,494 has been rounded down for an evaluation of 1,000,000 council tax dwellings in Wales, with each household using six 100-watt tungsten filament bulbs, then six 60-watt tungsten filament bulbs.

Taking the period of usage, as mentioned above, and assigning 30 days to each month (for simplification) we have a total of 180 days - it is assumed the lighting is on from 4.00pm to 11.00pm giving a total of 7 hours per days consumption.

Thus for one dwelling the total usage in hours for six months = 7 x 180 = 1260 hours

Using six 100-watt tungsten bulbs for 6 months

Total power used per dwelling = 1260 x 6 x 100 x 0.001 = **756 kWh**

Cost per dwelling at 12 pence per unit = 756 x 12 x o.o1 = **£90.72**

Comparison using six 20-watt CFL bulbs for 6 months

Total power used per dwelling = 1260 x 6 x 20 x 0.001 = **151.2 kWh**

Cost per dwelling at 12 pence per unit = 151.2 x 12 x 0.01 = **£18.14**

Six month units saved per dwelling = 756 – 151.2 = **604.8 kWh**

Six month cost saving per dwelling = £90.72 - £18.14 = **£72.58**

Annual saving per dwelling

Assuming each dwelling uses half the amount during the six summer months (April to September) then for the twelve months we have:

Power saving per dwelling = 604.8 + (604.8 x 0.5) = **907.2 kWh**

Cost saving per dwelling = £72.58 + (£72.58 x 0.5) = **£108.87**

Now changing six 100-watt tungsten light bulbs to six 60-watt tungsten light bulbs we have:

Using six 60-watt tungsten bulbs

Total power used per dwelling = 1260 x 6 x 60 x 0.001 = **453.6 kWh**

Cost per dwelling at 12 pence per unit = 453.6 x 12 x 0.01 = **£54.43**

Comparison using six 20-watt CFL bulbs

It is claimed a 20-watt CFL bulb will easily give the same level of lighting (brightness) as a 60-watt filament bulb.

Six month units saved per dwelling = 453.6 – 151.2 = **302.4 kWh**

Six month cost saving per dwelling = £54.43 - £18.14 = **£36.29**

Annual saving per dwelling

Power saving per dwelling = 302.4 + (302.4 x 0.5) = **453.6 kWh**

Cost saving per dwelling = £36.29 + (£36.29 x 0.5) = **£54.43**

The above calculations demonstrate quite clearly, that for a small country such as Wales, the amount of money that can be saved annually by changing from 60-watt tungsten bulbs to 20-watt CFL bulbs for domestic lighting alone amounts to a staggering **£54.43M** - the savings for a large country of say 20 million households would therefore be an eye-watering **£1088.6M** - and is easily achievable by simply moving to more effective technology such as CFLs.

Light Emitting Diodes (LEDs)

We will now explore the savings that can be made by moving to a MUCH more effective technology such as Light Emitting Diodes (LEDs) for domestic lighting.

Since 2010 the author has been assessing the Pharox 300 LED bulb that was produced by Lemnis Lighting (Technology Pioneer 2009 Award), The Netherlands. This marvellous bulb has a 90 per cent energy saving and a lifetime of 25 years. It is indeed a very impressive device demanding only 6 watts of electrical power.

The Pharox 300 LED bulb uses 90 per cent less power than an incandescent bulb, coupled with having many more advantages than the existing energy-saving light bulb (CFL) – such as being able to be used with a dimmer switch.

The following summarises the Pharox 300 benefits:

- Saves up to 90 per cent energy, compared to incandescent, up to 50 percent over CFL.

- Lifetime around 25 years, based on 4 hours of operation daily.

- Warm white light (+/- 3000K).

- Very energy efficient – 6 watt, replaces up to 60 watt incandescent.

- Dimmable.

- High light output (> 300 lumen).

- Fits most regular fixtures.

- Contains no mercury.

To demonstrate the efficiency a comparison is made between a 60-watt tungsten bulb and a 6-watt Pharox 300 LED bulb.

Taking the same period of usage for the CFL bulb as the six winter months October to March, assigning 30 days to each month to give a total of 180 days - it is assumed again that the lighting is on from 4.00pm to 11.00pm giving a total of 7 hours per day usage. Thus for a single dwelling the total usage in hours at 7 hours per day for six months = 7 x 180 = 1260 hours.

Using six 60-watt tungsten bulbs

Total power used per dwelling = **453.6 kWh**

Cost per dwelling at 12 pence per unit = **£54.43.**

Comparison using six 6-watt LED bulbs

Total LED power used per dwelling = 1260 x 6 x 6 x 0.001 = **45.36 kWh**

LED cost per dwelling at 12 pence per unit = 45.36 x 12 x 0.01 = **£5.44**

Six month units saved per dwelling = 453.6 – 45.36 = **408.34 kWh**

Six month cost saving per dwelling = £54.43 - £5.44 = **£49.**

Annual saving per dwelling

Again assuming the household uses half the winter lighting energy during the summer months, we have:

Power saving per dwelling = 408.34 + (408.34 x 0.5) = **612.5 kWh**

Cost saving per dwelling = 49 + (49 x 0.5) = **£73.5.**

Thus a potential saving for 1,000,000 households of **£73.5M** - a breath-taking **£1470M** for 20 million households

If the six 6-watt LED bulbs were to replace six 100-watt tungsten bulbs then we have:

Power saving per dwelling = 756 – 45.36 = **710.64 kWh**

Cost saving per dwelling = £90.72 - £5.44 = **£85.28.**

Again assuming the household uses half the winter lighting energy during the summer months then:

Annual saving per dwelling

Power saving per dwelling = 710.64 + (710.64 x 0.5) = **1066 kWh**

Cost saving per dwelling = £85.28 + (£85.28 x 0.5) = **£128**

Saving for 1,000,000 households of **£128M – an astounding £2560M for 20 million households.**

Summary

Although there are considerable savings in using CFL bulbs instead of the now obsolete incandescent bulbs, the final prize has to go to LED technology. The above exercises clearly demonstrate the effectiveness of the LED bulb and how much money can be saved - although the high cost (£30 each at 2010) of the Pharox 300 LED light bulb has to be considered such that 6 million Pharox bulbs at £30 each would mean a capital cost of £180M to purchase. Using the savings the LED bulb can achieve in a year for each dwelling (£73.5) then for a million dwellings after 2.5 years (180/73.5) the savings in energy consumption would have paid for the bulbs - and saving £73.5M per annum thereafter.

The Pharox 300 bulb has a lifetime of about 25 years. Therefore after a conservative 20 years then a total of £1470M (20 x 73.5) will be saved – but in reality the saving would be even greater - surely if the company that produces this bulb is given an order for 6 million bulbs, then economy of scale would kick in, and the cost per bulb would be far less – remembering that LED bulbs have dropped appreciably in price since 2010.

As an example, Welsh Government, instead of supporting ineffective wind farms and solar parks, could purchase a large quantity of LED bulbs and sell them cheaply to the public – indeed, why not give the bulbs free to every Welsh household and convert all public lighting to LED technology – this would result in a large reduction in electricity consumption, less emissions, on-going savings in electricity consumption (bulbs have at least 20-year life-span), and call a halt to all wind generator and solar park countryside industrialising development, thus protecting the tourist industry - everyone a winner as they say.

The Energy Saving Trust on their website (www.energysavingtrust.org.uk) state and I quote: "There are two main types of energy efficient light bulbs available in the UK. Compact Fluorescent Lamps (CFLs) and Light Emitting Diodes (LEDs).

CFLs are what you typically think of as an energy efficient light bulb. CFLs are a cost-effective option for most general lighting requirements. Replacing a traditional light bulb with a compact fluorescent light bulb (CFL) of the same brightness will save you around £3 per year, or £45 over the lifetime of the bulb.

LEDs are available to fit both types of fittings and are particularly good for replacing spotlights and dimmable lights. Though more expensive to buy initially, they are more efficient than CFLs and will save you more money in the long term. By replacing all halogen down-lighters in your home with LED alternatives, you could save around £40 a year on your electricity bills." End of quote.

Indeed if just two 100 watt bulbs were used for 4 hours of every day of a year, then the annual cost at 10 pence per unit would amount to: 2 x 100 x 4 x 365 x 0.1 x 0.001 = £29.20. Using 6 watt Pharox 300 LED bulbs then the annual cost would amount to: 2 x 6 x 4 x 365 x 0.1 x 0.001 =£1.752, thus giving an annual saving of: £29.20-£1.752 = **£27.448**.

The average power consumption in Wales per annum is of the order of 15 TWh, thus if we assume a unit (kWh) of electricity is 10 pence, we can calculate the saving achievable by using LEDs for all lighting:

Now 15 TWh = 15,000,000,000 kWh

If we assume that lighting uses approximately 25 per cent of all energy consumed we have:

15,000,000,000 x 0.25 = 3750,000,000 kWh

Thus if LEDs were employed at 90 per cent efficiency then the energy savings would be:

3750,000,000 x 0.9 = **3375,000,000 kWh**

A cost saving of 3375,000,000 x 0.1 = **£337.5M per annum at 10 pence per unit.**

The various examples above clearly show the savings achievable for lighting with LED technology.

Domestic Power Saving

The first example is a four bedroomed detached house in Pembrokeshire, from 1999 until 2013, and details the energy savings achievable from installing a solar PV system and using low energy lighting. The second example involves a five bedroomed detached house in Ceredigion, and considers not only a solar PV system and low energy lighting (predominantly LED), but air-to-air heating, voltage regulation and a Solar iBoost unit that has the ability to identify and enable the consumption of 'exported power solar energy' at source.

Four Bedroom Detached House, Pembrokeshire

Solar Power Installation in Pembrokeshire for the year 2012.

Latitude: 52.01492.   Longitude: -4.6176

Panels mounted on a detached garage roof with a tilt of approximately 35 degrees and a southerly orientation.

Solar array consists of 10 x 230 watt Dimplex high performance polycrystalline solar PV modules.

Total system capacity = 2.3 kW

Capital cost of installation £7818

Annual mains power consumption = 5000 kWh. This figure needs qualifying by noting a mix of low energy and LED technology being used for lighting, and gas (LPG) for heating and lighting.

The mean annual power consumption from the local power network from 1999 to 2013 was of the order of 5000 kWh per year, and the solar generation (2056.2 kWh) during 2012 turned out to be 40 per cent of the mains consumption - and this, during a predominantly cloudy year.

Solar Energy Earnings, Pembrokeshire

At the start of generation the FIT paid 0.433 pence per kWh generated and it was assumed, by the electricity supply company that 50 per cent of all generation would be fed back (exported) into the local network at a payment of 0.031 pence per kWh.

Total solar generation for 2012 = **2056.2 kWh.**

FIT payment 2012

Generation Tariff = 2056.2 x 0.433 = £890.33

Export Tariff = 2056.2 x 0.5 x 0.031 = 31.87

Total Tariff payment for 2012 = £890.33 + £31.87 = £922.2.

Note: For simplicity sake the initial value of the FIT is used, but please be aware that due to index linking the value of the FIT increased at April to 0.454 pence for each unit generated and 0.032 pence for each unit exported – using these figures would have made the savings marginally greater.

Electricity Bill Savings, Pembrokeshire

If the solar electricity generated is fully consumed at source:

The cost per unit (kWh) = 10 pence

Normal usage 5000 kWh

Reduced consumption due to solar = 5000 – 2056.2 = 2943.8 kWh
Reduced power bill due to solar generation = 2943.8 x 0.1 = £294.4

Therefore cost saving of £500 - £294.4 = **£205.6 per annum.**

Please note that the above saving is only true if ALL the solar panel generated power is used at source. This has not been the case as a certain amount of the solar generated power had in fact been exported to the local power network.

A good example of when export took place would be on a sunny day in the summer, when the occupants have spent the day at the seaside, leaving the house empty and thus a minimal need for electrical energy - low energy demanding items such as fridges and freezers - the property in Pembrokeshire contained a fridge in the kitchen and two freezers in the garage.

Bearing in mind that at other times of the year as during the winter months, when solar generation is low, most if not all, solar generation was used at source.

Total payments from the FIT amounted to £922.2 and the Electricity bill savings came to £205.6 giving a grand annual figure of **£1127.8 per annum for earnings and savings.**

20 Year projected earnings and savings, Pembrokeshire

Potential generation over 20 year period, 20 x 2056.2 = **41,124 kWh.**

Capital cost of system = £7818

Earnings and savings in year 1 = £1127.8

Therefore system pays for itself (7818/1127.8) = **6.9 years.**

Projecting over 20 years the system earns, 20 x £922.2 = **£18,444.**

Projecting over 20 years the system saves on mains electricity, 20 x £205.6 = **£4,112.**

Thus potential total earnings and savings over 20 year period, £18,444 + £4112 = **£22,556**

Deducting cost of system at £7818, potential total earnings and savings =**£14,738.**

SUMMARY

When the author purchased the solar PV system for the house in Pembrokeshire installation costs were high – but this was countered by the good FIT payments offered at the time which offered a very good investment on capital expenditure coupled with an attractive saving on the electricity bill – not forgetting being environmentally friendly.

Although the FIT payments are that much lower today, system installation is also that much cheaper - having reduced in price by about 50 per cent - a 2.5 kW system should now be obtainable for much less than £4000 – indeed, I have seen 4 kW installations advertised for less than £4000 - so solar PV systems are still a very attractive proposition, as we shall see now for Ceredigion.

Five Bedroom Detached Property, Ceredigion

During 2013 the author moved to a five bedroomed detached property in Ceredigion and demonstrated that even greater savings could be achieved in energy usage by employing not only solar PV panels, and low energy lighting, but air-to-air heating, voltage regulation and an electronic unit known as a **SOLAR iBOOST** which has the ability to identify 'exported solar energy' and intelligently control and adjust the flow of this energy to the house immersion heater instead of the local power network.

With the correct approach and a basic understanding of electrical power most property owners should have little difficulty in reducing their energy consumption - without any meaningful reduction to their quality of life – something every householder would wish to achieve in this age of ever increasing energy costs.

Solar Power Installation, Ceredigion

Latitude: 52.09884   Longitude: -4.6170

Panels mounted on the property roof with a tilt of approximately 40 degrees and a south westerly orientation.

Solar array consists of 16 x Hyundai 250 Watt Black Frame Polycrystalline panels

Total system capacity = 4 kW

Capital cost of installation = £7100

Annual mains power consumption for 2013, without solar PV panels, was assessed at 6000 kWh – this assessment being based on a five bedroomed house and the use of conventional/incandescent light bulbs.

The actual mains power consumption for 2014 was 4200 kWh and is the result of predominantly LED lighting, and the installation of solar PV system.

Heating by oil (underfloor and radiator) with an annual consumption of about 2000 litres.

Cooking by a combined electrical and gas (LPG) stove - with LPG gas consumption approximately 38 kg per annum - it should be noted that some back-up heating is obtained from a wall fitted gas fire in the lounge during extremely cold weather.

Total solar generation for 2014 = **4131 kWh.**

FIT payment 2014

Total solar generation = 4131 kWh

Generation Tariff = 4131 x 0.149 = £615.5

Export Tariff = 4131 x 0.5 x 0.0464 = £95.8

Total Tariff payment for 2014 = 615.5 + 95.8 = £711.

Note for simplicity sake (ignoring Index Linking) the prior April Tariff figures of 14.9 pence per kWh for generation and 4.64 pence per kWh for exporting of power have been used.

Electricity Bill Savings, Ceredigion

The cost per unit (kWh) = 12.28 pence

Typical amount of electricity used per annum = 6000 kWh

Typical power bill = 6000 x 12.28 x 0.1 = £736.89

Actual mains electricity for consumed for 2014 = 4200 kWh

Cost of electricity used = 4200 x 12.28 x 0.1 = £515.76

Therefore cost saving of £736.89 - £515.76 = **£221 per annum.**

Total solar power generation for 2014 = 4131 kWh.

Projected generation over 20 years = 20 x 4131 = 82,620 kWh

Total Annual Earnings and Savings, Ceredigion

Total annual earnings from FIT were £711.3 and the Electricity Bill savings were £221 giving an annual figure of 711 + 221 = **£932.**

20 Year projected earnings and savings, Ceredigion

Capital cost of system = £7100

Potential generation over 20 year period, 20 x 4131 = **82,620 kWh.**

Earnings and savings in year 1 = £932

System pays for itself (7100/932) = **7.6 years.**

Projecting over 20 years the system earns, 20 x 711.3 = **£14,226.**

Projecting over 20 years the system saves, 20 x 221 = **£4,420.**

Thus potential earnings and savings over 20 year period, £14226 + £4420 = **£18,646.**

Deducting cost of system at £7100, potential earnings and savings = **£11546.**

Summary

Site	Capital Cost of Installation £	Payback Time (years)	Potential Generation over 20 Years (kWh)	Potential FIT Earnings over 20 Years (£)	Actual Annual Solar Generation (kWh)
Pembrokeshire	7,818	6.9	41,124	18,444	2056
Ceredigion	7,100	7.6	82,620	14,226	4131

Note:

- ❖ Payback time is shorter for Pembrokeshire.

- ❖ A difference of £718 between the capital cost of the two sites.

- ❖ Potential generation at Ceredigion over 20 years is just over twice that of Pembrokeshire.

- ❖ Pembrokeshire potentially earns £4218 more than Ceredigion over a 20 year period.

- ❖ Ceredigion (4 kW) generation capacity is 1.7 kW greater than that of Pembrokeshire (2.3 kW).

- ❖ Actual solar generation for Ceredigion is double that of Pembrokeshire.

Although Ceredigion potential generation over 20 years is approximately double that of Pembrokeshire for the same period, the potential earnings are not. Predominantly this due to the smaller FIT payment (14.9 pence per kWh for Ceredigion compared to 43.3 pence per kWh for Pembrokeshire) and to a lesser degree the difference in solar energy generated at the two sites:

Pembrokeshire = (20 x 2056.2 x 0.433) + (20 x 2056.2 x 0.5 x 0.031) = (17,806.7) + (637.422) = £18,444.

Ceredigion = (20 x 4131 x 0.149) + (20 x 4131 x 0.5 x 0.0464) = (12310) + (1916) = £14,226.

Nevertheless even with the reduction in FIT payment the installation of solar PV panels is still regarded as a very good investment for the householder, especially when adding in the savings achieved by using less mains electricity.

If the solar PV system is enhanced by additional energy saving devices such as an air-to-air pump then a considerable saving in the cost of heating can be achieved – and if a device known as a Solar iBoost unit is installed then greater savings again can be achieved with the household hot water

Solar iBoost

Most solar PV systems are installed with just a simple generation meter which does not differentiate between generation used at source and the exported generation to the local power network - the installed meter can only show the total solar generation. As such it should be understood that the householder will not be aware of energy exported at any one moment and from the householders point of view this energy will be lost to the power network – recognising this energy being available to another paying consumer connected to the network.

This is where the Solar iBoost comes into its own as the unit has ability to identify 'exported solar energy' and intelligently control and divert this flow of energy to the immersion heater instead of the local power network. In effect this means that all of the power generated by a solar PV system can be used at source, much to the delight of the householder…a considerable saving in cost for heating the hot water tank over time.

A Solar iBoost unit was installed at the house in Ceredigion at the beginning of October 2014 and at the end of March, 2015 (6 months) had diverted 370 kWh of electrical energy to the immersion heater of the house hot water tank, thereby saving in the use of oil for heating hot water – considering the time of year this was an extremely satisfying outcome – at a domestic unit (kWh) price of 12.28 pence per unit for electricity this is a saving of £45.44 for the six winter months when days are predominantly cloudy and the Sun is low in the sky.

As the days lengthen toward the summer solstice, the savings will be far greater - toward the autumnal equinox the savings will still be substantial – and thus back to the six winter months when the cycle begins again - please note the savings in domestic oil were not assessed over the winter period due to lack of any suitable measuring equipment. But in conjunction with other energy saving devices there has been a considerable saving in household oil usage that has been quantified as we will see later when considering the air-to air heat energy saving device.

Summary of Energy Savings in the Property at Ceredigion for 2014

- ✓ Reduced mains consumption of electricity as a result of solar panels = 4200 kWh.
- ✓ Savings on cost of mains electricity £221.
- ✓ Generated 4131 kWh electricity from solar panels.
- ✓ Earned a total of £711 from the FIT payments.
- ✓ Total earnings and savings from solar panels £932.

Savings from Air-to-Air Pump

- ✓ Saved approximately 300 litres of heating oil.
- ✓ Cost saving in heating oil at 54.66 pence per litre £164.
- ✓ Household also enjoyed higher level of comfort due to higher levels of heating.

Voltage Regulator

Not able to assess power saved but company claims

- ✓ Lighting = 22 per cent.
- ✓ Generic Power Outlets = 7 per cent.
- ✓ Heating, ventilation and Aircon = 12 per cent.
- ✓ Kitchen appliances = 14 per cent.

Solar iBoost and Voltage Regulator 2014 - 2015

The unit was fitted at the beginning of October 2014

- ✓ Diverted 370 kWh to immersion heater up until the end of March 2015 (six winter months).

- ✓ Electricity savings at 12.28 pence per unit (kWh) = £45.44.

- ✓ Saving in oil usage for heating hot water.

Total earnings and saving for Ceredigion, 2014, coupled with an increased level of (heating) comfort.

£932 + £164 + £45.44 = **£1141.44 Annual saving.**

It should be noted that if the Solar iBoost unit can deliver 370 kWh to the immersion heater during the winter months when the Sun is at its lowest in the sky and the weather is predominantly cloudy, then it is not rocket science to realise that from April until October (British Summer Time) the unit will easily provide all the hot water necessary for the household at Ceredigion providing a saving in both electricity and oil costs. Therefore during the summer months there will be no oil usage for heating or hot water – the solar panels, air-to-air heater and Solar iBoost will easily cater for all household demands.

So what is there to dislike especially when earning and saving over £1100 per annum.

Appendix 4

Energy Statistics

Source: Wales.gov.uk and Department of Energy & Climate Change.

Energy can be generated from various sources of natural fuel, such as gas, coal and oil.

However energy generation usually refers specifically to electricity generation, which is the process of generating electric power from sources of energy. This section looks at the amount of electricity generated in Wales on an annual basis and how this has changed over time.

The standard approach to measuring electricity generation on a national scale is in gigawatt hours (GWh).

Prior to 2008 the amount of electricity generated in Wales remained relatively stable with around 35,000 GWh generated each year. However in recent years the amount of electricity generated in Wales has been falling, with 27,300 GWh generated in 2011.

It is noticeable that this change in trend occurred during a time when the country was in an economic downturn.

Across the UK as a whole and amongst the devolved administrations electricity generation has also been generally falling in recent years, with the exception of Scotland. This may be due to reduced demand, possibly as a result of the economic climate, introduction of energy efficiency measures or milder winters.

There has been a considerable decrease in consumption (11 per cent in Wales between 2005 and 2010). During 2011 the electricity generated in Wales accounted for just over 7 per cent of all the electricity generated in the UK.

Source: DECC Sub-National Electricity Consumption and www.wales.gov.uk/statistics.

	Annual Generation (GWh)	Annual Consumption (GWh)
2009	31,988	15,720
2010	32,170	15,818
2011	27,300	15,226
2012	26,201	13,524

The 2008 Living in Wales property survey estimated that around a fifth of households in Wales used a main heating fuel other than mains gas. This is particularly prevalent in rural local authorities which have a much higher percentage of households with no gas connection.

To take account of this issue the domestic consumption information is presented per consumer so that comparisons can be made between different local authorities.

In 2011, the valleys authority of Blaenau Gwent had the highest domestic gas consumption per consumer (15,200 kWh) but the lowest domestic electricity consumption per consumer (3,300 kWh). Conversely, Gwynedd experienced the lowest domestic gas consumption per consumer (12,100 kWh) but had one of the highest electricity consumption per consumer (4,600 kWh) figures.

Over the last year, domestic gas consumption per consumer fell across all local authorities. The largest percentage decrease was seen in Newport (9.2 per cent) followed by Denbighshire and Ceredigion (both 8.9 per cent).

All local authorities also saw a fall in domestic electricity consumption per consumer, with the exception of Bridgend which increased slightly (0.2 per cent). The largest fall was in Ceredigion (4.4 per cent) although it continued to have the highest average domestic electricity consumption per consumer (5,000 kWh) amongst all local authorities.

Over the last year, domestic gas consumption per consumer fell across all local authorities.

The largest percentage decrease was seen in Newport (9.2 per cent) followed by Denbighshire and Ceredigion (both 8.9 per cent).

All local authorities also saw a fall in domestic electricity consumption per consumer, with the exception of Bridgend which increased slightly (0.2 per cent). The largest fall was in Ceredigion (4.4 per cent) although it continued to have the highest average domestic electricity consumption per consumer (5,000 kWh) amongst all local authorities.

Wales Local Authority Average Domestic Electricity per Customer, data from wales.gov.uk.

2010 = 3,916 kWh

2011 = 3,845 kWh

Wales is a net exporter of the electricity it generates, which differs to England which imports electricity from Wales, Scotland and from continental Europe.

This means that Wales exports electricity generated here to consumers elsewhere in Great Britain (GB). This is because Wales generally has more generation capacity than it requires whilst England generally has less.

Generally the amount of electricity exported from Wales follows the same trend as the amount of electricity generated in Wales, with both the amount of electricity exported and generated falling since peaking in 2008.

Between 2010 and 2011 exported electricity fell by 54 per cent to a record low of 3,700 GWh. During 2011, the amount of electricity exported from Wales was equivalent to 13 per cent of total electricity generated in Wales. Previously exports have accounted for between 18 and 30 per cent of all electricity generated in a given year.

UK Electricity Generation and Supply (GWh), Data from DECC.

	2011	2012	2013
Total Generation (excl. pumped storage)	364,346	360,439	356,253
Total Supply	373,473	375,277	373,581

Average Electricity Consumption per Householder for 2013 (DECC).

4,170 kWh.

Average Gas Consumption per Household for 2013, data from (DECC).

14,829 kWh.

Glossary of Terms

AC. Alternating current.

Alternating Current. An electrical current that reverses direction at a regular rate. In the United Kingdom this happens at fifty times a second and is known the frequency; see also Frequency and Hertz.

Aerial. A conductor or arrangement of conductors radiating or collecting electromagnetic energy.

Ammeter. An instrument for measuring the amount of current in an electrical circuit.

Amperage. The quantity of electrical current flowing in an electrical circuit.

Ampere. Unit of current. A current of 1 ampere represents the movement of 6,280,000,000,000,000,000 electrons past a given point in a circuit during 1 second of time.

Anode. The positive terminal of a cell or battery. In a cathode-ray tube or electronic valve the anode is the plate to which a positive voltage is applied.

Armature. The moving part or parts of an electrical motor or generator - also the moving part of a relay, bell or buzzer.

Atom. The smallest particle of an element that has all the element's chemical properties, composed of a nucleus and a number of surrounding electrons.

Atomic nucleus. The central part of an atom consisting of protons and neutrons. The protons have a positive charge, giving the nucleus a positive electric charge; the neutrons have no electric charge.

AWEA. American Wind Energy Association.

Bond. Usually refers to a conducting bond by which the lead sheath and the armour of one or more cables, or the casing or framework of electrical apparatus/machinery are electrically connected together and/or to earth.

Battery. A group of electrical cells connected in series or parallel.

BERR. Department of Business, Enterprise and Regulatory Reform, formerly DTI.

BNFL. British Nuclear Fuels.

Brushes. Devices that provide stationary connections to the rotor in an electrical generator or motor. Carbon is commonly used for brushes in electrical hand tools.

Bulb. See lamp.

BWEA. British Wind Energy Association.

Cable. An arrangement of stranded conductors to form one common core/conductor and insulated throughout its length. A number of cores may be enclosed in a protective sheathing – which may be further protected by armour.

Capacity factor. American terminology for load factor, see also load factor.

Cathode. The negative terminal of a cell or battery. Also the plate/source in a thermionic device from which electrons are emitted.

Cathode-ray tube. An electronic device that converts electrical signals to a visual display on a fluorescent screen. See also CRT.

CCGT. Abbreviation for Combined-cycle gas turbine.

Cell. A source of electrical energy dependent upon chemical action. A voltaic cell is made of two different kinds of conductor materials placed within a paste or fluid (called an electrolyte) that also conducts electricity.

Coil. Turns of wire conductor to concentrate a magnetic field.

Commutator. A device used to reverse the direction of an electrical current. A generator or motor commutator is made of insulated copper bars, or segments mounted on a shaft. These bars are connected to the coils of wire wound into the armature core.

Circuit. An arrangement of conductors and components connected together to carry an electric current.

Circuit Breaker. A switch for making and breaking an electrical circuit under normal or fault conditions.

Co-generation. The generation of electrical energy and usable heat in the form of hot water or steam, from the same quantity of fuel in a single operation.

Conductor. A substance which offers low resistance to the flow of an electrical current: a solid, liquid or gas through which electrons can pass easily - gold, silver and copper are good electrical conductors.

Converter. A device for converting power from AC to DC and vice versa.

CRT. Cathode-ray tube.

Current. The flow of electricity around an electrical circuit.

DC. Direct current.

DECC. Department of Energy and Climate Change.

Direct Current. An electrical current that flows always in the same direction.

Distribution network. The system (low voltage) to which consumer services are connected.

DTI. The old Department of Trade and Industry.

Dukes. Digest of UK energy statistics.

Earthed Circuit. An electrical circuit in which one or more points are connected to earth. See also Ground.

Efficiency. The ratio of energy output to energy input – usually expressed as a percentage.

EDF. Electricité de France (EDF Energy).

Electricity, Dynamic. A form of energy present when electrons move through a circuit.

Electricity, Static. A form of energy present within the space between two oppositely charged objects.

Electricity Meter. An instrument in the consumers premises which totals up the electrical energy supplied over a given time.

Electric polarisation. The separation of charge in an object so that one part bears a positive charge whilst another part bears an equal negative charge.

Electrode. A conducting body employed to pass an electric current into and out of an electrolyte, gas or electronic valve/tube.

Electrolysis. The process of decomposing an electrolyte.

Electrolyte. A solution through which the passage of an electric current will cause chemical decomposition.

Electromagnet. A magnet produced by an electric current moving through a coil wound around a core of iron.

Electromagnetism. The magnetism produced by an electric current.

Electromagnetic Induction. The production of an electromotive force by changing magnetic fields.

Electron. A negatively charged particle in the shell of an atom; electrons spin about the nucleus of an atom.

Electrostatic Field. The space around a charged body where the lines of electric force may be detected.

E.M.F. Short for electromotive force. It is the electrical force that causes electrons to move through an electrical conductor/circuit and is commonly known as voltage.

E.on UK. Subsidiary of E.ON Nnetz. (Formerly Powergen).

E.ON Nnetz. German based power company.

EWEA. European Wind Energy Association.

Frequency. The number of cycles an alternating current completes over a period of time. Used to be expressed in cycles per second but now known as Hertz – one Hertz equals one cycle per second.

Fuse. A protective device to give protection to a circuit or circuits against excessive currents.

Generator. A machine that converts mechanical energy to electrical energy.

Gigawatt. One thousand Megawatts. Usually abbreviated to GW.

Grid. The network of high-voltage transmission lines.

Ground. Also known as Earth – an electrical connection/path between an electrical circuit and the earth.

Hertz. The unit of frequency abbreviated as Hz. One Hertz is equal to one cycle per second.

Hydroelectricity. The means by which electrical energy is generated by water turbines.

Insulator. A material which offers a high resistance to an electric current.

Kilovolt. One thousand volts. Usually abbreviated to kV.

Kilowatt. Unit of power. One thousand watts. Usually abbreviated to kW.

Kilowatt-hour. Unit of electrical energy. Usually abbreviated to kWh.

Lamp. A device for converting electrical energy to light energy.

LCD. Liquid-crystal display.

LED. Light-emitting diode.

Load. The electrical power carried by a circuit or taken from a generator; the amount of electricity used by a device when connected to an electrical supply.

Load Factor. The actual amount of electricity produced by a generator/s compared to the maximum amount possible over the same period of time, and expressed as a percentage. Also known as Capacity Factor.

Magnetic Field. The space surrounding a magnet or current carrying conductor, where magnetic lines of force may be detected.

Magnetic Flux. The number of magnetic lines of force passing through a given cross-section.

Megawatt. One thousand kilowatts. Usually abbreviated to MW.

Nuclear Power. The means by which electrical energy is produced by nuclear reaction; heat from a nuclear reactor is used to produce steam to drive a steam turbine.

Ohm. The unit of electrical resistance.

Power. The amount of electrical energy delivered in a unit of time.

Power Station. One or more large electrical generators housed in a large building.

Primary Cell. A chemical cell that cannot be recharged.

Rectifier. A device that allows current to flow in one direction only and is used to convert alternating current into direct current.

Resistance. The opposition to current flow through a conductor or circuit; the unit of resistance is the ohm.

Resonance. The state of a system in which the natural period of oscillation is the same as that of the impulses to which it is subjected.

Rotor. The rotating part of an AC machine.

Secondary Cell. Chemical cell that can be recharged.

Series Connection. Conductors, Resistances or circuits are said to be in series when they are connected so that the same current flows in each conductor, resistance or circuit.

Sub-station. A point in an electrical supply area at which electricity is supplied in bulk. The electricity is then, via switchgear, transformers and cables directed to suit the system of supply to the particular area. It is important to note that sub-stations do not generate electricity.

Switch. A device in an electrical circuit that allows (closed) or stops (open) the flow of an electrical current.

Synchronism. The condition when two machines or two systems have the same frequency and are in phase.

Synchroscope. An instrument used to determine the phase relationship between two alternating voltages of the same frequency.

Terawatt. One thousand Gigawatts. Usually abbreviated to TW.

Transformer. A device employing electromagnetic induction to transform alternating power in one winding (primary winding) to alternating power in another winding (secondary winding) usually at different values of current and voltage.

Transmission. The supply of electrical energy at high voltages, usually 132 kV and higher.

Transmission Line. Power lines carried on tall towers called pylons to supply electrical energy at high voltages, see also transmission.

Volt. Unit of electromotive force or potential difference.

Voltage. The electromotive force between different points in an electrical circuit measured in volts.

Voltmeter. An instrument used to measure voltage.

Watt. Unit of power usually abbreviated to W.

OTHER BOOKS BY THE AUTHOR

PAPERBACK BOOK

COUNTDOWN TO OBLIVION
The definitive Alien abduction

Obtainable from Amazon

Or Trafford Publishing
ISBN 141202685-7
www.trafford.com
E-mail sales: sales@trafford.com

KINDLE cartoon books

Obtainable from Amazon

I just love my computer

Computer Rage

Romans, Greeks, Egyptians and Gauls

Aliens and Space

KINDLE fantasy book for children

Troll Castle and the Forbidden Chamber of Gold

The Author

The author is a retired telecommunications engineering manager, who was employed for almost forty years with a large telecommunications organisation - during the course of his career he gained theoretical and field experience in electrical disciplines such as lightning protection, and rise-of-earth-potential protection at sub-stations and large power stations including the pumped storage system at Dinorwig, Trawsfynydd Nuclear Power Station (now de-commissioned), Llanelli Power station, which had some of the first wind generators on site, Ironbridge fossil fuelled power station, and many others.

Instrumental in developing and introducing a deep driven earth electrode for earth electrode systems, with experience in introducing lightning protection for underground and overhead telecommunications cables.

Examination qualifications consist of a National Certificate in Electrical Engineering and a Full Technological Certificate (Telecommunications) which included a distinction in digital computing at year five; many years ago a member of the Institute of Electrical and Electronic Technician Engineers (IEETE).

The author exploring the volcanic landscape in Tenerife with Mount Teide in the background

Hobbies include amateur astronomy (member of the Shropshire Astronomical Society), gardening, reading, DIY, travelling, computing, amateur radio (certificate to qualify transmitting and receiving on FM only), a passionate love and appreciation of coastal and country walking.

During his very early years and as a member of the Youth Hostels Association (YHA) the author with two friends cycled from Cardiff, via paddle steamer across the Bristol Channel to Weston Super Mare, then on to Land's End and back over a period of a week, staying at various Youth Hostels along the route which included cycling over Dartmoor and Bodmin Moor – years later he completed the YHA (7 peaks) 40 mile marathon walk which entailed climbing the seven highest peaks in South Wales, starting at 0500hrs from the Llanddeusant Youth Hostel in the Black Mountains, Carmarthenshire, crossing and climbing the Brecon Beacons, and completing the marathon the same day at approximately 2200hrs at the George VI Memorial Youth Hostel, Capel-y-Ffin, Black Mountains near Hay-on-Wye.

The author had a power boat moored in the Teifi Estuary for a number of years, and when at sea, he and his wife enjoyed the beautiful Welsh coastal scenery, mackerel and pollock fishing, seal watching and being enchanted by the inquisitive approach and playful 'showing off' by the dolphins in Cardigan Bay.

www.ingramcontent.com/pod-product-compliance
Lightning Source LLC
Chambersburg PA
CBHW031612210526
45464CB00004B/1536